Cognitive Surplus

Also by Clay Shirky

Here Comes Everybody:
The Power of Organizing Without Organizations

Cognitive Surplus

Creativity and Generosity in a Connected Age

CLAY SHIRKY

THE PENGUIN PRESS
New York
2010

THE PENGUIN PRESS
Published by the Penguin Group
Penguin Group (USA) Inc., 375 Hudson Street, New York,
New York 10014, U.S.A. • Penguin Group (Canada),
90 Eglinton Avenue East, Suite 700, Toronto, Ontario,
Canada M4P 2Y3 (a division of Pearson Penguin Canada Inc.) •
Penguin Books Ltd, 80 Strand, London WC2R 0RL, England •
Penguin Ireland, 25 St. Stephen's Green, Dublin 2, Ireland
(a division of Penguin Books Ltd) • Penguin Books Australia Ltd,
250 Camberwell Road, Camberwell, Victoria 3124, Australia
(a division of Pearson Australia Group Pty Ltd) • Penguin Books
India Pvt Ltd, 11 Community Centre, Panchsheel Park,
New Delhi – 110 017, India • Penguin Group (NZ),
67 Apollo Drive, Rosedale, North Shore 0632, New Zealand
(a division of Pearson New Zealand Ltd) •
Penguin Books (South Africa) (Pty) Ltd, 24 Sturdee Avenue,
Rosebank, Johannesburg 2196, South Africa

Penguin Books Ltd, Registered Offices:
80 Strand, London WC2R 0RL, England

First published in 2010 by The Penguin Press,
a member of Penguin Group (USA) Inc.

LIBRARY OF CONGRESS CATALOGING-IN-PUBLICATION DATA
Shirky, Clay.
Cognitive surplus : creativity and generosity in a connected age /
by Clay Shirky.
p. cm.
Includes bibliographical references and index.
ISBN 978-1-59420-253-7
1. Information society. 2. Social media. 3. Mass media—Social aspects.
I. Title.
HM851.S5464 2010
303.48'33—dc22 2009053882

Printed in the United States of America
1 3 5 7 9 10 8 6 4 2

DESIGNED BY AMANDA DEWEY

For Red Burns

Contents

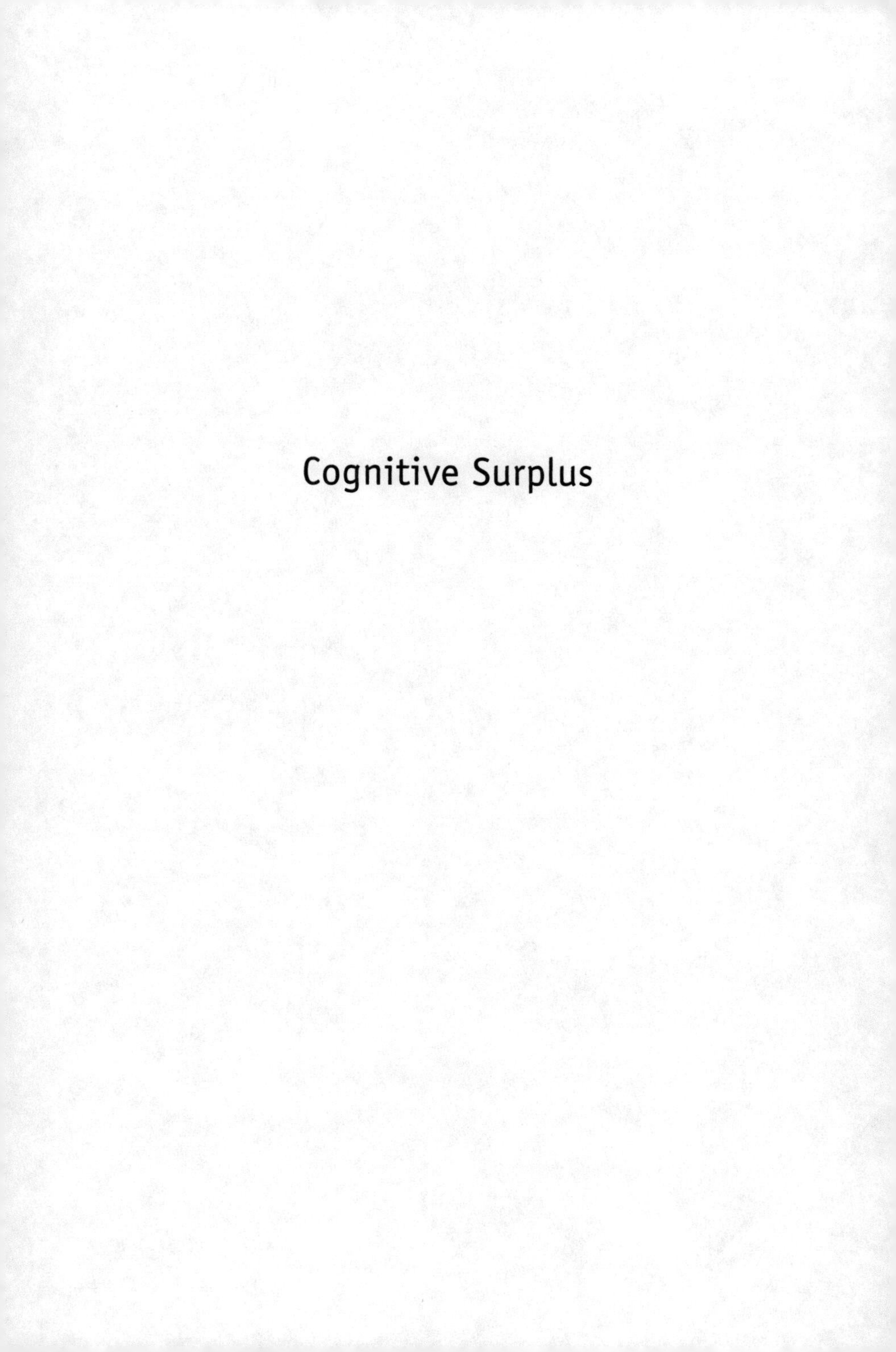

Cognitive Surplus

1

Gin, Television, and Cognitive Surplus

In the 1720s, London was busy getting drunk. Really drunk. The city was in the grips of a gin-drinking binge, largely driven by new arrivals from the countryside in search of work. The characteristics of gin were attractive: fermented with grain that could be bought locally, packing a kick greater than that of beer, and considerably less expensive than imported wine, gin became a kind of anesthetic for the burgeoning population enduring profound new stresses of urban life. These stresses generated new behaviors, including what came to be called the Gin Craze.

Gin pushcarts plied the streets of London; if you couldn't afford a whole glass, you could buy a gin-soaked rag, and flophouses did brisk business renting straw pallets by the hour if you needed to sleep off the effects. It was a kind of social lubricant for people suddenly tipped into an unfamiliar and often unforgiving life, keeping them from completely falling apart. Gin offered its consumers the ability to fall apart a little bit at a time. It was a collective bender, at civic scale.

The Gin Craze was a real event—gin consumption rose dramatically in the early 1700s, even as consumption of beer and wine remained flat. It was also a change in perception. England's wealthy and titled were increasingly alarmed by what they saw in the streets of London. The population was growing at a historically unprecedented rate, with predictable effects on living conditions and public health, and crime of all sorts was on the rise. Especially upsetting was that the women of London had taken to drinking gin, often gathering in mixed-sex gin halls, proof positive of its corrosive effects on social norms.

It isn't hard to figure out why people were drinking gin. It is palatable and intoxicating, a winning combination, especially when a chaotic world can make sobriety seem overrated. Gin drinking provided a coping mechanism for people suddenly thrown together in the early decades of the industrial age, making it an urban phenomenon, especially concentrated in London. London was the site of the biggest influx of population as a result of industrialization. From the mid-1600s to the mid-1700s, the population of London grew two and a half times as fast as the overall population of England; by 1750, one English citizen in ten lived there, up from one in twenty-five a century earlier.

Industrialization didn't just create new ways of working, it created new ways of living, because the relocation of the population destroyed ancient habits common to country living, while drawing so many people together that the new density of the population broke the older urban models as well. In an attempt to restore London's preindustrial norms, Parliament seized on gin. Starting in the late 1720s, and continuing over the next three decades, it passed law after law prohibiting various aspects of gin's production, consumption, or sale. This strategy was ineffective, to put it mildly. The result was instead a thirty-year cat-and-mouse game of legislation to prevent gin consumption, followed by the rapid invention of ways to defeat those laws. Parliament outlawed "flavored spirits"; so distillers stopped adding juniper berries to the liquor. Selling gin was made illegal; women sold from bottles hidden beneath their skirts, and some entrepreneurial types created the "puss and mew," a cabinet set on the streets where a customer could approach and, if they knew the password, hand their money to the vendor hidden inside and receive a dram of gin in return.

What made the craze subside wasn't any set of laws. Gin consumption was treated as the problem to be solved, when it fact it was a reaction to the real problem—dramatic social change and the inability of older civic models to adapt. What helped the Gin Craze subside was the restructuring of society around the new urban realities created by London's incredible social density, a restructuring that turned London into what we'd recognize as a modern city, one of the first. Many of the institutions we mean when we talk about "the industrialized world" actually arose in response to the social climate created by industrialization, rather than to industrialization itself. Mutual aid societies provided shared management of risk outside the traditional ties of kin and church. The spread of coffeehouses and later restaurants was spurred by concentrated populations. Political parties began to recruit the urban poor and to field candidates more responsive to them. These changes came about only when civic density stopped being treated as a crisis and started being treated as a simple fact, even an opportunity. Gin consumption, driven upward in part by people anesthetizing themselves against the horrors of city life, started falling, in part because the new social structures mitigated these horrors. The increase in both population and aggregate wealth made it possible to invent new kinds of institutions; instead of madding crowds, the architects of the new society saw a civic surplus, created as a side effect of industrialization.

And what of us? What of our historical generation? That section of the global population we still sometimes refer to as "the industrialized world" has actually been transitioning to a postindustrial form for some time. The postwar trends of emptying rural populations, urban growth, and increased suburban density, accompanied by rising educational attainment across almost all demographic groups, have marked a huge increase in the number of people paid to think or talk, rather than to produce or transport objects. During this transition, what has been our gin, the critical lubricant that eased our transition from one kind of society to another?

The sitcom. Watching sitcoms—and soap operas, costume dramas, and the host of other amusements offered by TV—has absorbed the lion's share of the free time available to the citizens of the developed world.

Since the Second World War, increases in GDP, educational attainment, and life span have forced the industrialized world to grapple with something we'd never had to deal with on a national scale: free time. The amount of unstructured time cumulatively available to the educated population ballooned, both because the educated population itself ballooned, and because that population was living longer while working less. (Segments of the population experienced an upsurge of education and free time before the 1940s, but they tended to be in urban enclaves, and the Great Depression reversed many of the existing trends for both schooling and time off from work.) This change was accompanied by a weakening of traditional uses of that free time as a result of suburbanization—moving out of cities and living far from neighbors—and of periodic relocation as people moved for jobs. The cumulative free time in the postwar United States began to add up to billions of collective hours per year, even as picnics and bowling leagues faded into the past. So what did we do with all that time? Mostly, we watched TV.

We watched *I Love Lucy*. We watched *Gilligan's Island*. We watched *Malcolm in the Middle*. We watched *Desperate Housewives*. We had so much free time to burn and so few other appealing ways to burn it that every citizen in the developed world took to watching television as if it were a duty. TV quickly took up the largest chunk of our free time: an average of over twenty hours a week, worldwide. In the history of media, only radio has been as omnipresent, and much radio listening accompanies other activities, like work or travel. For most people most of the time, watching TV *is* the activity. (Because TV goes in through the eyes as well as the ears, it immobilizes even moderately attentive users, freezing them on chairs and couches, as a prerequisite for consumption.)

The sitcom has been our gin, an infinitely expandable response to the crisis of social transformation, and as with drinking gin, it isn't hard to explain why people watch individual television programs—some of them are quite good. What's hard to explain is how, in the space of a generation, watching television became a part-time job for every citizen in the developed world. Toxicologists like to say "The dose makes the poison"; both alcohol and caffeine are fine in moderation but fatal in excess. Similarly, the question of TV isn't about the content of individual shows but about their volume:

the effect on individuals, and on the culture as a whole, comes from the dose. We didn't just watch good TV or bad TV, we watched everything—sitcoms, soap operas, infomercials, the Home Shopping Network. The decision to watch TV often preceded any concern about what might be on at any given moment. It isn't what we watch, but how much of it, hour after hour, day after day, year in and year out, over our lifetimes. Someone born in 1960 has watched something like fifty thousand hours of TV already, and may watch another thirty thousand hours before she dies.

This isn't just an American phenomenon. Since the 1950s, any country with rising GDP has invariably seen a reordering of human affairs; in the whole of the developed world, the three most common activities are now work, sleep, and watching TV. All this is despite considerable evidence that watching that much television is an actual source of unhappiness. In an evocatively titled 2007 study from the *Journal of Economic Psychology*—"Does Watching TV Make Us Happy?"—the behavioral economists Bruno Frey, Christine Benesch, and Alois Stutzer conclude that not only do unhappy people watch considerably more TV than happy people, but TV watching also pushes aside other activities that are less immediately engaging but can produce longer-term satisfaction. Spending many hours watching TV, on the other hand, is linked to higher material aspirations and to raised anxiety.

The thought that watching all that TV may not be good for us has hardly been unspoken. For the last half century, media critics have been wringing their hands until their palms chafed over the effects of television on society, from Newton Minow's famous description of TV as a "vast wasteland" to epithets like "idiot box" and "boob tube" to Roald Dahl's wicked characterization of the television-obsessed Mike Teavee in *Charlie and the Chocolate Factory*. Despite their vitriol, these complaints have been utterly ineffective—in every year of the last fifty, television watching per capita has grown. We've known about the effects of TV on happiness, first anecdotally and later through psychological research, for decades, but that hasn't curtailed its growth as the dominant use of our free time. Why?

For the same reason that the disapproval of Parliament didn't reduce the consumption of gin: the dramatic growth in TV viewing wasn't the problem, it was the reaction to the problem. Humans are social creatures, but the explosion of our surplus of free time coincided with a steady reduction in social capital—our stock of relationships with people we trust and rely on. One clue about the astonishing rise of TV-watching time comes from its displacement of other activities, especially social activities. As Jib Fowles notes in *Why Viewers Watch*, "Television viewing has come to displace principally (a) other diversions, (b) socializing, and (c) sleep." One source of television's negative effects has been the reduction in the amount of human contact, an idea called the social surrogacy hypothesis.

Social surrogacy has two parts. Fowles expresses the first—we have historically watched so much TV that it displaces all other uses of free time, including time with friends and family. The other is that the people we see on television constitute a set of imaginary friends. The psychologists Jaye Derrick and Shira Gabriel of the University at Buffalo and Kurt Hugenberg of Miami University of Ohio concluded that people turn to favored programs when they are feeling lonely, and that they feel less lonely when they are viewing those programs. This shift helps explain how TV became our most embraced optional activity, even at a dose that both correlates with and can cause unhappiness: whatever its disadvantages, it's better than feeling like you're alone, even if you actually are. Because watching TV is something you can do alone, while it assuages the feelings of loneliness, it had the right characteristics to become popular as society spread out from dense cities and tightly knit rural communities to the relative disconnection of commuter work patterns and frequent relocations. Once a home has a TV, there is no added cost to watching an additional hour.

Watching TV thus creates something of a treadmill. As Luigino Bruni and Luca Stanca note in "Watching Alone," a recent paper in the *Journal of Economic Behavior and Organization*, television viewing plays a key role in crowding-out social activities with solitary ones. Marco Gui and Luca Stanca take on the same phenomenon in their 2009 working paper "Television Viewing, Satisfaction and Happiness": "television can play a significant role in raising people's materialism and material aspirations, thus leading individuals to underestimate the relative importance of interpersonal relations for their life satisfaction and, as a consequence, to over-invest in income-producing activities and under-invest in relational activities." Translated from the dry language of economics, *under-investing in relational activities* means spending less time with friends and family, precisely because watching a lot of TV leads us to shift more energy to material satisfaction and less to social satisfaction.

Our cumulative decision to commit the largest chunk of our free time to consuming a single medium really hit home for me in 2008, after the publication of *Here Comes Everybody*, a book I'd written about social media. A TV producer who was trying to decide whether I should come on her show to discuss the book asked me, "What interesting uses of social media are you seeing now?"

I told her about Wikipedia, the collaboratively created encyclopedia, and about the Wikipedia article on Pluto. Back in 2006, Pluto was getting kicked out of the planet club—astronomers had concluded that it wasn't enough like the other planets to make the cut, so they proposed redefining *planet* in such a way as to exclude it. As a result, Wikipedia's Pluto page saw a sudden spike in activity. People furiously edited the article to take account of the proposed change in Pluto's status, and the most committed group of editors disagreed with one another about how best to characterize the change. During this conversation, they updated the article—contesting sections, sentences, and even word choice throughout—transforming the essence of the article from "Pluto is the ninth planet" to "Pluto is an odd-shaped rock with an odd-shaped orbit at the edge of the solar system."

I assumed that the producer and I would jump into a conversation about social construction of knowledge, the nature of authority, or any of the other topics that Wikipedia often generates. She didn't ask any of those questions, though. Instead, she sighed and said, "Where do people find the time?" Hearing this, I snapped, and said, "No one who works in TV gets to ask that question. You know where the time comes from." She knew, because she worked in the industry that had been burning off the lion's share of our free time for the last fifty years.

Imagine treating the free time of the world's educated citizenry as an aggregate, a kind of cognitive surplus. How big would that surplus be? To figure it out, we need a unit of measurement, so let's start with Wikipedia. Suppose we consider the total amount of time people have spent on it as a kind of unit—every edit made to every article, and every argument about those edits, for every language that Wikipedia exists in. That would represent something like one hundred million hours of human thought, back when I was talking to the TV producer. (Martin Wattenberg, an IBM researcher who has spent time studying Wikipedia, helped me arrive at that figure. It's a back-of-the-envelope calculation, but it's the right order of magnitude.) One hundred million hours of cumulative thought is obviously a lot. How much is it, though, compared to the amount of time we spend watching television?

Americans watch roughly two hundred *billion* hours of TV every year. That represents about two thousand Wikipedias' projects' worth of free time annually. Even tiny subsets of this time are enormous: we spend roughly a hundred million hours every weekend just watching commercials. This is a pretty big surplus. People who ask "Where do they find the time?" about those who work on Wikipedia don't understand how tiny that entire project is, relative to the aggregate free time we all possess. One thing that makes the current age remarkable is that we can now treat free time as a general social asset that can be harnessed for large, communally created projects, rather than as a set of individual minutes to be whiled away one person at a time.

Society never really knows what to do with any surplus at first. (That's what makes it a surplus.) For most of the time when we've had a truly large-scale surplus in free time—billions and then trillions of hours a year—we've spent it consuming television, because we judged that use of time to be better than the available alternatives. Sure, we could have played outdoors or read books or made music with our friends, but we mostly didn't, because the thresholds to those activities were too high, compared to just sitting and watching. Life in the developed world includes a lot of passive participation: at work we're office drones, at home we're couch potatoes. The pattern is easy enough to explain by assuming we've wanted to be passive participants more than we wanted other things. This story has been, in the last several decades, pretty plausible; a lot of evidence certainly supported this view, and not a lot contradicted it.

But now, for the first time in the history of television, some cohorts of young people are watching TV less than their elders. Several population studies—of high school students, broadband users, YouTube users—have noticed the change, and their basic observation is always the same: young populations with access to fast, interactive media are shifting their behavior away from media that presupposes pure consumption. Even when they watch video online, seemingly a pure analog to TV, they have opportunities to comment on the material, to share it with their friends, to label, rate, or rank it, and of course, to discuss it with other viewers around the world. As Dan Hill noted in a much-cited online essay, "Why *Lost* Is Genuinely New Media," the viewers of that show weren't just viewers—they collaboratively created a compendium of material related to that show called (what else?) Lostpedia. Even when they are engaged in watching TV, in other words, many members of the networked population are engaged with one another, and this engagement correlates with behaviors other than passive consumption.

The choices leading to reduced TV consumption are at once tiny and enormous. The tiny choices are individual; someone simply decides to spend the next hour talking to friends or playing a game or creating something instead of just watching. The enormous choices are collective ones, an accumulation of those tiny choices by the millions; the cumulative shift toward participation across a whole population enables the creation of a Wikipedia. The television industry has been shocked to see alternative uses of free time, especially among young people, because the idea that watching TV was the best use of free time, as ratified by the viewers, has been such a stable feature of society for so long. (Charlie Leadbeater, the U.K. scholar of collaborative work, reports that a TV executive recently told him that participatory behavior among the young will go away when they grow up, because work will so exhaust them that they won't be able to do anything with their free time but "slump in front of the TV.") Believing that the past stability of this behavior meant it would be a stable behavior in the future as well turned out to be a mistake—and not just any mistake, but a particular kind of mistake.

MILKSHAKE MISTAKES

When McDonald's wanted to improve sales of its milkshakes, it hired researchers to figure out what characteristics its customers cared about. Should the shakes be thicker? Sweeter? Colder? Almost all of the researchers focused on the product. But one of them, Gerald Berstell, chose to ignore the shakes themselves and study the customers instead. He sat in a McDonald's for eighteen hours one day, observing who bought milkshakes and at what time. One surprising discovery was that many milkshakes were purchased early in the day—odd, as consuming a shake at eight A.M. plainly doesn't fit the bacon-and-eggs model of breakfast. Berstell also garnered three other behavioral clues from the morning milkshake crowd: the buyers were always alone, they rarely bought anything besides a shake, and they never consumed the shakes in the store.

The breakfast-shake drinkers were clearly commuters, intending to drink them while driving to work. This behavior was readily apparent, but the other researchers had missed it because it didn't fit the normal way of thinking about either milkshakes or breakfast. As Berstell and his colleagues noted in "Finding the Right Job for Your Product," their essay in the *Harvard Business Review,* the key to understanding what was going on was to stop viewing the product in isolation and to give up traditional notions of the morning meal. Berstell instead focused on a single, simple question: "What job is a customer hiring that milkshake to do at eight A.M.?"

If you want to eat while you are driving, you need something you can eat with one hand. It shouldn't be too hot, too messy, or too greasy. It should also be moderately tasty, and take a while to finish. Not one conventional breakfast item fits that bill, and so without regard for the sacred traditions of the morning meal, those customers were hiring the milkshake to do the job they needed done.

All the researchers except Berstell missed this fact, because they made two kinds of mistakes, things we might call "milkshake mistakes." The first was to concentrate mainly on the product and assume that everything important about it was somehow implicit in its attributes, without regard to what role the customers wanted it to play—the job they were hiring the milkshake for.

The second mistake was to adopt a narrow view of the type of food people have always eaten in the morning, as if all habits were deeply rooted traditions instead of accumulated accidents.

Neither the shake itself nor the history of breakfast mattered as much as customers needing food to do a nontraditional job—serve as sustenance and amusement for their morning commute—for which they hired the milkshake.

We have the same problems thinking about media. When we talk about the effects of the web or text messages, it's easy to make a milkshake mistake and focus on the tools themselves. (I speak from personal experience—much of the work I did in the 1990s focused obsessively on the capabilities of computers and the internet, with too little regard for the way human desires shaped them.)

The social uses of our new media tools have been a big surprise, in part because the possibility of these uses wasn't implicit in the tools themselves. A whole generation had grown up with personal technology, from the portable radio through the PC, so it was natural to expect them to put the new media tools to personal use as well. But the use of a social technology is much less determined by the tool itself; when we use a network, the most important asset we get is access to one another. We want to be connected to one another, a desire that the social surrogate of television deflects, but one that our use of social media actually engages.

It's also easy to assume that the world as it currently exists represents some sort of ideal expression of society, and that all deviations from this sacred tradition are both shocking and bad. Although the internet is already forty years old, and the web half that age, some people are still astonished that individual members of society, previously happy to spend most of their free time consuming, would start voluntarily making and sharing things. This making-and-sharing is certainly a surprise compared to the previous behavior. But pure consumption of media was never a sacred tradition; it was just a set of accumulated accidents, accidents that are being undone as people start hiring new communications tools to do jobs older media simply can't do.

To pick one example, a service called Ushahidi was developed to help citizens track outbreaks of ethnic violence in Kenya. In December 2007 a disputed election pitted supporters and opponents of President Mwai Kibaki against one another. Ory Okolloh, a Kenyan political activist, blogged about the violence when the Kenyan government banned the mainstream media from reporting on it. She then asked her readers to e-mail or post comments about the violence they were witnessing on her blog. The method proved so popular that her blog, Kenyan Pundit, became a critical source of first-person reporting. The observations kept flooding in, and within a couple of days Okolloh could no longer keep up with it. She imagined a service, which she dubbed Ushahidi (Swahili for "witness" or "testimony"), that would automatically aggregate citizen reporting (she had been doing it by hand), with the added value of locating the reported attacks on a map in near-real time. She floated the idea on her blog, which attracted the attention of the programmers Erik Hersman and David Kobia. The three of them got on a conference call and hashed out how such a service might work, and within three days, the first version of Ushahidi went live.

People normally find out about the kind of violence that took place after the Kenyan election only if it happens nearby. There is no public source where people can go to locate trouble spots, either to understand what's going on or to offer help. We've typically relied on governments or professional media to inform us about collective violence, but in Kenya in early 2008 the professionals weren't covering it, out of partisan fervor or censorship, and the government had no incentive to report anything.

Ushahidi was developed to aggregate this available but dispersed knowledge, to collectively weave together all the piecemeal awareness among individual witnesses into a nationwide picture. Even if the information the public wanted existed someplace in the government, Ushahidi was animated by the idea that rebuilding it from scratch, with citizen input, was easier than trying to get it from the authorities. The project started as a website, but the Ushahidi developers quickly added the ability to submit information via text message from mobile phones, and that's when the reports really poured in. Several months after Ushahidi launched, Harvard's Kennedy School of Government did an analysis that compared the site's data to that of the mainstream media and concluded that Ushahidi had been better at reporting acts of violence as they started, as opposed to after the fact, better at reporting acts of nonfatal violence, which are often a precursor to deaths, and better at reporting over a wide geographical area, including rural districts.

All of this information was useful—governments the world over act less violently toward their citizens when they are being observed, and Kenyan NGOs used the data to target humanitarian responses. But that was just the beginning. Realizing the site's potential, the founders decided to turn Ushahidi into a platform so that anyone could set up their own service for collecting and mapping information reported via text message. The idea of making it easy to tap various kinds of collective knowledge has spread from the original Kenyan context. Since its debut in early 2008,

Ushahidi has been used to track similar acts of violence in the Democratic Republic of Congo, to monitor polling places and prevent voter fraud in India and Mexico, to record supplies of vital medicines in several East African countries, and to locate the injured after the Haitian and Chilean earthquakes.

A handful of people, working with cheap tools and little time or money to spare, managed to carve out enough collective goodwill from the community to create a resource that no one could have imagined even five years ago. Like all good stories, the story of Ushahidi holds several different lessons: People want to do something to make the world a better place. They will help when they are invited to. Access to cheap, flexible tools removes many of the barriers to trying new things. You don't need fancy computers to harness cognitive surplus; simple phones are enough. But one of the most important lessons is this: once you've figured out how to tap the surplus in a way that people care about, others can replicate your technique, over and over, around the world.

Ushahidi.com, designed to help a distressed population in a difficult time, is remarkable, but not all new communications tools are so civically engaged; in fact, most aren't. For every remarkable project like Ushahidi or Wikipedia, there are countless pieces of throwaway work, created with little effort, and targeting no positive effect greater than crude humor. The canonical example at present is the lolcat, a cute picture of a cat that is made even cuter by the addition of a cute caption, the ideal effect of "cat plus caption" being to make the viewer laugh out loud (thus putting the *lol* in *lolcat*). The largest collection of such images is a website called ICanHasCheezburger.com, named after its inaugural image: a gray cat, mouth open, staring maniacally, bearing the caption "I

Can Has Cheezburger?" (Lolcats are notoriously poor spellers.) ICanHasCheezburger.com has more than three thousand lolcat images—"i have bad day," "im steelin som ur foodz k thx bai," "BANDIT CAT JUST ATED UR BURRITOZ"—each of which garners dozens or hundreds of comments, also written in lolspeak. We are far from Ushahidi now.

Let's nominate the process of making a lolcat as the stupidest possible creative act. (There are other candidates, of course, but lolcats will do as a general case.) Formed quickly and with a minimum of craft, the average lolcat image has the social value of a whoopee cushion and the cultural life span of a mayfly. Yet anyone seeing a lolcat gets a second, related message: *You can play this game too*. Precisely because lolcats are so transparently created, anyone can add a dopey caption to an image of a cute cat (or dog, or hamster, or walrus—Cheezburger is an equal-opportunity time waster) and then share that creation with the world.

Lolcat images, dumb as they are, have internally consistent rules, everything from "Captions should be spelled phonetically" to "The lettering should use a sans-serif font." Even at the stipulated depths of stupidity, in other words, there are ways to do a lolcat wrong, which means there are ways to do it right, which means there is some metric of quality, even if limited. However little the world needs the next lolcat, the message *You can play this game too* is a change from what we're used to in the media landscape. The stupidest possible creative act is still a creative act.

Much of the objection to lolcats focuses on how stupid they are; even a funny lolcat doesn't amount to much. On the spectrum of creative work, the difference between the mediocre and the good is vast. Mediocrity is, however, still on the spectrum; you can move from mediocre to good in increments. The real gap is between doing nothing and doing something, and someone making lolcats has bridged that gap.

As long as the assumed purpose of media is to allow ordinary people to consume professionally created material, the proliferation of amateur-created stuff will seem incomprehensible. What amateurs do is so, well, unprofessional—lolcats as a kind of low-grade substitute for the Cartoon Network. But what if, all this time, providing professional content isn't the only job we've been hiring media to do? What if we've also been hiring it to make us feel connected, engaged, or just less lonely? What if we've always wanted to produce as well as consume, but no one offered us that opportunity? The pleasure in *You can play this game too* isn't just in the making, it's also in the sharing. The phrase "user-generated content," the current label for creative acts by amateurs, really describes not just personal but also social acts. Lolcats aren't just user-generated, they are user-shared. The sharing, in fact, is what makes the making fun—no one would create a lolcat to keep for themselves.

The atomization of social life in the twentieth century left us so far removed from participatory culture that when it came back, we needed the phrase "participatory culture" to describe it. Before the twentieth century, we didn't really have a phrase for participatory culture; in fact, it would have been something of a tautology. A significant chunk of culture was participatory—local gatherings, events, and performances—because where else could culture come from? The simple act of creating something with others in mind and then sharing it with them represents, at the very least, an echo of that older model of culture, now in technological raiment. Once you accept the idea that we actually like making and sharing things, however dopey in content or poor in execution, and that making one another laugh is a different kind of activity from being made to laugh by people paid to make us laugh, then in some ways the Cartoon Network is a low-grade substitute for lolcats.

MORE IS DIFFERENT

When one is surveying a new cultural effusion like Wikipedia or Ushahidi or lolcats, answering the question *Where do people find the time?* is surprisingly easy. We have always found the time to do things that interest us, specifically because they interest us, a resource fought for in the struggle to create the forty-hour workweek. Amid the late-nineteenth-century protests for better working conditions, one popular workers' chant was "Eight hours for work, eight hours for sleep, eight hours for what we will!" For more than a century now, the explicit and specific availability of unstructured time has been part of the bargain of industrialization. Over the last fifty years, however, we've spent the largest chunk of that hard-won time on a single activity, a behavior so universal we've forgotten that our free time has always been ours to do with as we like.

People asking *Where do people find the time?* aren't usually looking for an answer; the question is rhetorical and indicates that the speaker thinks certain activities are stupid. In my conversation with the TV producer, I also mentioned *World of Warcraft*, an online game set in a fantasy realm of knights and elves and evil demons. Many of the challenges in *Warcraft* are so difficult that they cannot be undertaken by individual players; instead, the players have to band together into guilds, complex, in-game social structures with dozens of members, each performing specialized tasks. As I described these guilds and the work they require of their members, I could tell what she thought of *Warcraft* players: grown men and women sitting in their basements pretending to be elves? Losers.

The obvious response is: at least they're doing something.

Did you ever see that episode of *Gilligan's Island* where they almost get off the island and then Gilligan messes up and they don't? I saw that one a lot when I was growing up. And every half hour I watched it was a half hour in which I wasn't sharing photos or uploading video or conversing on a mailing list. I had an ironclad excuse—none of those things could be done in my youth, when I was committing my thousand hours a year to Gilligan and the Partridge Family and Charlie's Angels. However pathetic you may think it is to sit in your basement pretending to be an elf, I can tell you from personal experience: it's worse to sit in your basement trying to decide whether Ginger or Mary Ann is cuter.

Dave Hickey, the iconoclastic art historian and cultural critic, wrote an essay in 1997 called "Romancing the Looky-Loos," in which he talked about the varieties of audiences for music. The title of the essay comes from hearing his father, a musician, describe a particular audience as looky-loos, people there only to consume. To be a looky-loo is to approach an event, especially a live event, as if you were mindlessly watching it on TV: "They paid their dollar at the door, but they contributed nothing to the occasion—afforded no confirmation or denial that you could work with or around or against."

Participants are different. To participate is to act as if your presence matters, as if, when you see something or hear something, your response is part of the event. Hickey quotes musician Waylon Jennings discussing what it's like to perform for an audience that participates: "They seek you out in little clubs because they understand what you're doing, so you feel like you're doing it for them. And if you go wrong in these clubs, you know it immediately." Participants give feedback; looky-loos don't. The participation can happen well after the event—for whole communities of people, movies, books, and TV shows create more than an opportunity to consume; they create an opportunity to reply and discuss and argue and create.

Media in the twentieth century was run as a single event: consumption. The animating question of media in that era was *If we produce more, will you consume more?* The answer to that question has generally been yes, as the average person consumed more TV with each passing year. But media is actually like a triathlon, with three different events: people like to consume, but they also like to produce, and to share. We've always enjoyed all three of those activities, but until recently, broadcast media rewarded only one of them.

TV is unbalanced—if I own a TV station, and you own a television, I can speak to you, but you can't speak to me. Phones, by contrast, are balanced; if you buy the means of consumption, you automatically own the means of production. When you purchase a phone, no one asks if you just want to listen, or if you want to talk on it too. Participation is inherent in the phone, and it's the same for the computer. When you buy a machine that lets you consume digital content, you also buy a machine to produce it. Further, you can share material with your friends, and you can talk about what you consumed or produced or shared. These aren't additional features; they are part of the basic package.

Evidence accumulates daily that if you offer people the opportunity to produce and to share, they'll sometimes take you up on it, even if they've never behaved that way before and even if they're not as good at it as the pros. That doesn't mean we'll stop mindlessly watching TV. It just means that consumption will no longer be the only way we use media. And any shift, however minor, in the way we use a trillion hours of free time a year is likely to be a big deal.

Expanding our focus to include producing and sharing doesn't even require making big shifts in individual behavior to create enormous changes in outcome. The world's cognitive surplus is so large that small changes can have huge ramifications in aggregate. Imagine that everything stays 99 percent the same, that people continue to consume 99 percent of the television they used to, but 1 percent of that time gets carved out for producing and sharing. The connected population still watches well over a trillion hours of TV a year; 1 percent of that is more than one hundred Wikipedias' worth of participation per year.

Scale is a big part of the story, because the surplus has to be accessible in aggregate; for things like Ushahidi to work, people must be able to donate their free time to collective efforts and produce a cognitive surplus, instead of making just a bunch of tiny, disconnected individual efforts. Part of the story of aggregate scale has to do with how the educated population uses its free time, but another part of it has to do with the aggregation itself, with our being increasingly connected together in a single, shared media landscape. In 2010 the global internet-connected population will cross two billion people, and mobile phone accounts already number over three billion. Since there are something like 4.5 billion adults worldwide (roughly 30 percent of the global population is under fifteen), we live, for the first time in history, in a world where being part of a globally interconnected group is the normal case for most citizens.

Scale makes big surpluses function differently from small ones. I first discovered this principle three decades ago, when my parents sent me to New York City to visit a cousin for my sixteenth birthday. My reaction was pretty much what you'd expect from a midwestern kid dumped into that environment—awe at the buildings and the crowds and the hustle—but in addition to all the big things, I noticed a small one, and it changed my sense of the possible: pizza by the slice.

Where I grew up, I worked at a pizza chain called Ken's. Here I learned this: A customer asks for a pizza. You make a pizza. Twenty minutes later you hand that pizza to that customer. It was simple and predictable. But pizza by the slice isn't like that at all. You never know who is going to want a slice, yet you have to make a pie in advance, as the whole point for the customer is to be in and out in considerably less than twenty minutes, with a much smaller bit of pizza than an entire pie.

The meaning of pizza by the slice, the meaning that hit me at sixteen, is that with a large enough crowd, unpredictable events become predictable. On any given day, you no longer have to know who will want pizza to be certain that someone will want pizza, and once the certainty of demand is divorced from the individual customers and remanded to the aggregate, whole new classes of activity become possible. (If my sixteen-year-old self had had more working capital, I would have discovered the same principle by observing the logic of hailing a cab versus waiting at a bus stop.) More generally, the likelihood of an event is the probability of it happening times the frequency with which it might happen. Where I grew up, the chance that someone would want a single slice of pizza at three in the afternoon was too low to take a chance on. At the corner of Thirty-fourth Street and Sixth Avenue, on the other hand, you could build a whole business on those odds. Any human event, however improbable, sees its likelihood grow in a crowd. Big surpluses are different from small ones.

In the words of the physicist Philip Anderson, "More is different." When you aggregate a lot of something, it behaves in new ways, and our new communications tools are aggregating our individual ability to create and share, at unprecedented levels of more. Consider this question, one whose answer has changed dramatically in recent years: What are the chances that a person with a camera will come across an event of global significance? If you extrapolate your answer from an egocentric view—What are the chances I will witness such an event?—they are slim, indeed vanishingly small. And extrapolating from the personal chance can make the overall chance seem unlikely as well.

One reason we have such a hard time thinking about cultural change as enabled by new communications tools is that the egocentric view is the wrong way to approach it. The chance that anyone with a camera will come across an event of global significance is simply the number of witnesses of the event times the percentage of them that have cameras. That first number will fluctuate up and down depending on the event, but the second number—the number of people carrying cameras—rose from a few million worldwide in 2000 to well over a billion today. Cameras are now embedded in phones, increasing the numbers of people who have a camera with them all the time.

We've seen the effects of this new reality dozens of times: the London transport bombings in 2005, the Thai coup in 2006, the police killing of Oscar Grant in Oakland in 2008, the post-election Iranian unrest in 2009—all these events and countless more were documented with camera phones and then uploaded for the world to see. The chance that someone with a camera will come across an event of global significance is rapidly becoming the chance that such an event has any witnesses at all. Those kinds of changes in scale mean that formerly improbable events become likely, and that formerly unlikely events become certainties. Where we previously relied on professional photojournalists to document such events, we are increasingly becoming one another's infrastructure. This may be a cold-blooded way of looking at sharing—that we increasingly learn about the world through strangers' random choices about what to share—but even that has some human benefit. As Kurt Vonnegut's protagonist says at the close of *The Sirens of Titan*, "The worst thing that could possibly happen to anybody would be to not be used for anything by anybody." The ways in which we are combining our cognitive surplus make that fate less likely by the day.

Because we are increasingly producing and sharing media, we have to relearn what that word can mean. The simple sense of *media* is the middle layer in any communication, whether it is as ancient as the alphabet or as recent as mobile phones. On top of this straightforward and relatively neutral definition is another notion, inherited from the patterns of media consumption over the last several decades, that media refers to a collection of businesses, from newspapers and magazines to radio and television, that have particular ways of producing material and particular ways of making money. And as long as we use media to refer just to those businesses, and to that material, the word will be an anachronism, a bad fit for what's happening today. Our ability to balance consumption with production and sharing, our ability to connect with one another, is transforming the sense of media from a particular sector of the economy to a cheap and globally available tool for organized sharing.

A NEW RESOURCE

This book is about the novel resource that has appeared as the world's cumulative free time is addressed in aggregate. The two most important transitions allowing us access to this resource have already happened—the buildup of well over a trillion hours of free time each year on the part of the world's educated population, and the invention and spread of public media that enable ordinary citizens, previously locked out, to pool that free time in pursuit of activities they like or care about. Those two facts are common to every story in this book, from inspirational work like Ushahidi to mere self-amusement like lolcats. Understanding those two changes, as different as they are from the media landscape of the twentieth century, is just the beginning of understanding what is happening today, and what is possible tomorrow.

My previous book, *Here Comes Everybody*, was about the rise of social media as a historical fact, and the changed circumstances for group action that appeared with it. This book picks up where that one left off, starting with the observation that the wiring of humanity lets us treat free time as a shared global resource, and lets us design new kinds of participation and sharing that take advantage of that resource. Our cognitive surplus is only potential; it doesn't mean anything or do anything by itself. To understand what we can make of this new resource, we have to understand not just the kind of actions it makes possible but the hows and wheres of those actions.

When the police want to understand whether someone could have taken a particular action, they look for means, motive, and opportunity. Means and motive are the how and why of a particular action, and opportunity is the where and with whom. Do people have the capability to do something with their cumulative free time, the motivation to do it, and the opportunity to do it? Positive answers to these questions help establish the link between the person and the action; expressed at a larger scale, accounts of means, motive, and opportunity can help explain the appearance of new behaviors in society. Understanding what our cognitive surplus is making possible means understanding the means by which we are aggregating our free time; our motivations in taking advantage of this new resource; and the nature of the opportunities that are being created, and that we are creating for each other, in fact. The next three chapters detail these *hows, whys,* and *whats* behind cognitive surplus.

Even that, though, doesn't yet describe what we could do with the cognitive surplus, because the way we put our collective talents to work is a social issue, not solely a personal one. Because we have to coordinate with one another to get anything out of our shared free time and talents, using cognitive surplus isn't just about accumulating individual preferences. The culture of the various groups of users matters enormously for what they expect of one another and how they work together. The culture in turn will determine how much of the value that we get out of the cognitive surplus will be merely communal (enjoyed by the participants, but not of much use for society at large) and how much of it will be civic. (You can think of *communal* versus *civic* as paralleling *lolcats* versus *Ushahidi*.) After I address means, motive, and opportunity in chapters 2–4, the subsequent two chapters take up these questions of user culture and of communal versus civic value.

The last chapter, the most speculative of all, details some of the lessons we've already learned from successful uses of cognitive surplus, lessons that can guide us as more of that surplus is used in more important ways. Because of the complexity of social systems generally, and especially of those with diverse, voluntary actors, no simple list of lessons can operate as a recipe, but they can serve as guide rails, helping keep new projects from running into certain difficulties.

The cognitive surplus, newly forged from previously disconnected islands of time and talent, is just raw material. To get any value out of it, we have to make it mean or do things. We, collectively, aren't just the source of the surplus; we are also the people designing its use, by our participation and by the things we expect of one another as we wrestle together with our new connectedness.

2

Means

Back in 2003, after several sources of beef in the United States were revealed to be contaminated with mad cow disease (technically known as bovine spongiform encephalopathy), South Korea banned American beef imports. That ban lasted, with small exceptions, for five years, and because South Korea had been the third-largest export market for U.S. beef, it became a significant sore point between the two governments. Finally, in April 2008, Presidents Lee Myung-bak and George W. Bush negotiated a re-opening of the Korean market to U.S. beef as a precursor to a much larger free trade agreement. This agreement ended the issue, or rather it seemed to, until the Korean public got involved.

In May of that year, as news broke that U.S. beef would return to the Korean market, Korean citizens staged public protests, turning out in Cheonggyecheon Park, a verdant tract running through the center of Seoul. The protests took the form of candlelight vigils, after which many stayed overnight in the park. These protests had several distinctive features, one of which was their longevity: rather than petering out, they lasted for several weeks. Then there was their sheer scale: though the demonstrations started small, they grew to thousands and ultimately tens of thousands. By early June, the protests were the largest in Korea since the 1987 protests that had ushered in the return of democratically elected government. So many people occupied Cheonggyecheon, for so long, that they killed large patches of grass.

Most unusual, though, were the protesters themselves, not just in number but in makeup. Korea's previous protests had mostly been organized by political or labor groups. But in the mad cow protests, over half the participants—including many of the earliest organizers—were teenagers, most notably teenage girls. These "candlelight girls" were too young to vote, they were not members of any political group, and most of them had not participated in public political action before. Their presence helped make the vigils Korea's first family-friendly protest; for over a month, whole families turned out in the park, often with young children and infants. When the world's governments survey the possible sources of national unrest, they don't usually worry about teenage girls. Where had they come from?

Those girls had always been there—they were, after all, Korean citizens—but they simply hadn't mobilized in large numbers before. Democracies both produce and rely upon complacency in their citizens. A democracy is working when its citizens are content enough not to turn out in the streets; when they do, it's a sign something isn't right. Seen this way, the girls' participation is a question of what had changed. What would cause girls too young to vote to turn out in the park, day after day and night after night, for weeks?

The South Korean government tried blaming political fringe actors and agents provocateurs bent on damaging its relationship with the United States, but the protests were so enormous and so long-lived that that explanation quickly rang hollow. How had those kids gotten radicalized? Mimi Ito, a cultural anthropologist at the University of Southern California who studies the intersection of teenage behavior and communications technologies, quoted a thirteen-year-old candlelight girl about her motivations: "I'm here because of Dong Bang Shin Ki."

Dong Bang Shin Ki isn't a political party or an activist organization. DBSK is a boy band (the name translates to "Rising Gods of the East"), and in the tradition of boy bands everywhere, each member embodies a type: there's Kim Junsu, the romantic cutie; Shim Changmin, tall, dark, and handsome; and so on. They are clean-cut and mostly apolitical, hardly important voices in matters of foreign policy or even in protest music. They are, however, a significant focal point for Korean girls. When the South Korean market was reopened to U.S. beef, the band's online fan site, Cassiopeia, had nearly a million users, and on one of those bulletin boards many of the protesters first heard of the ban being lifted.

I'm here because of Dong Bang Shin Ki isn't the same thing as *Dong Bang Shin Ki sent me;* DBSK never actually recommended any sort of public or even political involvement. Rather, its site provided these girls with an opportunity to discuss whatever they wanted, including politics. They had gotten upset—had upset one another, in fact—about both the health and political issues surrounding the reopening of the Korean market. Massed together, frightened and angry that Lee's government had agreed to what seemed a national humiliation and a threat to public health, the girls decided to do something about it.

DBSK's website provided a place and a reason for Korea's youth to gather together by the hundreds of thousands. Here the ephemeral conversations that take place in the schoolyard and the coffee shop acquired two features previously reserved for professional media makers: accessibility and permanence. Accessibility means that a number of others can read what a given person writes, and permanence refers to the longevity of a given bit of writing. Both accessibility and permanence are increased when people connect to the internet, and South Korea is the most connected nation on earth. The average Seoul resident has access to better, faster, and more widely available communications networks, both on their computers and on their mobile phones, than the average citizen of London, Paris, or New York.

Commercial media that covers DBSK, like the gossip sites Pop Seoul and K-Popped, would never have thought of asking their readers what they thought of the government's food-import policies. Like the gossip sites, the DBSK bulletin boards are not a specifically political environment, but unlike the gossip sites, they are not specifically apolitical either. They are shaped by their participants, taking on the characteristics that their participants want them to have. Mainstream Korean media reported on the lifting of the beef ban; a small number of professional media producers conveyed the information to a large number of mostly uncoordinated amateur media consumers (the normal pattern of broadcast and print media in the twentieth century). Whenever a DBSK fan posted anything on Cassiopeia, by contrast, whether it was about Kim's new haircut or the Korean government's import policies, it was as widely and publicly available as any article in a Korean newspaper, and more available than much of what was on TV (since anything on the web can be shared more easily than anything on TV). Furthermore, the recipients of these bits of amateur media weren't silent consumers but noisy producers themselves, able to both respond to and redistribute those messages at will. In the case of the mad cow protests, connected South Korean citizens, even thirteen-year-olds, radicalized one another.

It's not clear what South Korea's policy on U.S. beef should be. But the change Lee negotiated upset many citizens who wanted to be consulted and hadn't been. When kids who are too young to vote are out in the street protesting policies, it can shake governments used to a high degree of freedom from public oversight. In this case, the giant, continual protest around the hot-button issue of food safety (and, as the protest went on, education policy and national identity) eroded Lee's popularity. He had entered office in February 2008 with close to a 75 percent approval rating. But during the month of May, that figure plummeted to less than 20 percent.

As May turned to June, and the protesters didn't go away, Lee's government finally decided enough was enough and ordered police to break up the protest—a task it set about with gusto. Instantly, websites were filled with images of policemen with water cannons and batons attacking the largely peaceful protesters; thousands of people watched online videos of police clubbing or kicking teenage girls in the head. The crackdown had the opposite effect of the one Lee intended. Condemnation of the police was widespread, even international, and both the Asian Human Rights Commission and Amnesty International began investigations. As a result of the violence and subsequent publicity, the protest grew bigger.

June 10 is the anniversary of the end of South Korea's military government in the 1980s and the country's return to democracy. As that day approached in 2008, the demonstrations took on the feel of a general antigovernment protest. Running out of options, Lee went on national TV to apologize for lifting the ban without adequately consulting the Korean people, and for the way the protests had been broken up. He forced his entire cabinet to resign, negotiated additional restrictions on all beef imported from the United States, and explained to the citizens what was at stake for South Korea in the free trade agreement overall, saying to the public, "I was in a hurry after being elected president, as I thought I could not succeed unless I achieved changes and reform within one year after inauguration."

This strategy worked. Some groups were still dissatisfied with Lee, his government, and its specific policies, but hearing the president admit that he had made a mistake by not directly addressing the people, and seeing the mass firing of the cabinet, took the urgency out of the protests, which abated. Lee had won a partial victory, albeit at enormous political cost, but the groups in the park had also won something. The public wanted to be consulted on significant matters, and if that didn't happen through ordinary channels, places like the DBSK bulletin boards would provide all the coordination they needed.

In Seoul ordinary citizens used a communication medium that neither respects nor enforces silence among The People Formerly Known as the Audience, as my NYU colleague Jay Rosen likes to call us. We are used to the media's telling us things: the people on TV tell us that the South Korean government has banned U.S. beef because of fears of mad cow disease, or that it's lifted the ban.

During the protests in South Korea, though, media stopped being just a source of information and became a locus of coordination as well. Those kids in the park used the DBSK bulletin boards, as well as conversations on Daum, Naver, Cyworld, and a host of other conversational online spaces. They were also sending images and text via their mobile phones, not just to disseminate information and opinion but to act on it, both online and in the streets. In doing so, they changed the context in which the South Korean government operates.

The old view of online as a separate space, cyberspace, apart from the real world, was an accident of history. Back when the online population was tiny, most of the people you knew in your daily life weren't part of that population. Now that computers and increasingly computerlike phones have been broadly adopted, the whole notion of cyberspace is fading. Our social media tools aren't an alternative to real life, they are part of it. In particular, they are increasingly the coordinating tools for events in the physical world, as in Cheonggyecheon Park.

It's not clear what the longer-term effects of this increased public participation will be. The South Korean presidency runs for one five-year term, so Lee will never face the voters again. Moreover, the South Korean government is aggressively trying to require citizens to use their real names online. (Significantly, this restriction is only for sites with more than one hundred thousand visitors a month, giving the policy a distinctly political feel.) It is attempting to restore the populace to a state we might call forced complacency. The competition between the government and the people has thus become an arms race, but one that involves a new class of participants. When teenage girls can help organize events that unnerve national governments, without needing professional organizations or organizers to get the ball rolling, we are in new territory. As Ito describes the protesters,

> Their participation in the protests was grounded less in the concrete conditions of their everyday lives, and more in their solidarity with a shared media fandom. . . . Although so much of what kids are doing online may look trivial and frivolous, what they are doing is building the capacity to connect, to communicate, and ultimately, to mobilize. From Pokémon to massive political protests, what's distinctive about this historical moment and today's rising generation is not only a distinct form of media expression, but how this expression is tied to social action.

People concerned about digital media often worry about the decay of face-to-face contact, but in Seoul, the most wired (and wireless) place on earth, the effect was just the opposite. Digital tools were critical to coordinating human contact and real-world activity. The old idea that media is a domain relatively separate from "the real world" no longer applies to situations like the mad cow protests, or indeed to any of the myriad ways people are using social media to arrange real-world action. Not only is social media in a new set of hands—ours—but when communications tools are in new hands, they take on new characteristics.

PRESERVING OLD PROBLEMS

One practical problem that can now be taken on in a social way is transportation, especially commuting. Getting to and from work requires a significant effort, and billions undertake it five days a week. This problem doesn't seem at first glance to be related to media, but one of the principal solutions available to commuting is carpooling, and the key to carpooling isn't cars, it's coordination. Carpooling doesn't require new cars, just new information about existing ones.

PickupPal.com is one of those new information channels, a carpooling site designed to coordinate drivers and riders planning to travel along the same route. The driver proposes a price for the ride, and if the passenger agrees, the system puts them in touch with each other. As with many a one-sentence business plan, a million details lie under this one's hood, from figuring out how closely a route and time have to overlap to constitute an acceptable match, to putting drivers and passengers in touch with each other without disclosing too many personal details.

PickupPal also faces the problem of scale—below a certain threshold number of potential drivers and riders, the system will hardly work at all, while above that threshold more is better. Someone using the system and finding a match one time out of three will have a very different attitude toward it than someone who finds a match nine times out of ten. One in three is a backup plan; nine in ten is infrastructure. PickupPal's basic approach to the scale problem is to start where the potential for social coordination is high and to work outward from there. Since the system is most effective for commutes around big cities, PickupPal works with corporations and organizations, who can advertise carpooling opportunities to their employees or members (a strategy that also helps foster trust among users). It also integrates with existing social tools like Facebook in order to make finding other people as simple as possible. Taken together, these strategies seem to be working: at the end of 2009, PickupPal.com had more than 140,000 users in 107 countries.

The service PickupPal provides parallels our cognitive surplus in general. When each person has to solve the commuting problem entirely on their own, the solution is each person owning and driving their own car. But this "solution" makes the problem worse. Once we see the problem of commuting as a matter of coordination, however, we can think of aggregate solutions rather than just individual ones. In the context of carpooling, the number of cars on the road becomes an opportunity, because each additional car is an additional chance that someone will be going your way. PickupPal reimagines the surplus of cars and drivers as a potentially shared resource. As long as everyone has access to a medium that allows communication among groups, we can configure new approaches to transportation problems that rely on moving information around between drivers and riders, solutions that benefit almost everyone.

Almost everyone, but not bus companies. In May 2008 the Ontario-based bus company Trentway-Wagar hired a private detective to use PickupPal; the detective confirmed that it worked as advertised and produced an affidavit stating that he'd gotten a ride to Montreal, for which he'd reimbursed the driver sixty dollars. With this evidence, Trentway-Wagar then petitioned the Ontario Highway Transport Board (OHTB) to shut PickupPal down on the grounds that, by helping coordinate drivers and riders, it worked too well to be a carpool. Trentway-Wagar invoked Section 11 of the Ontario Public Vehicles Act, which stipulated that carpooling could happen only between home and work (rather than, say, school or hospital.) It had to happen within municipal lines. It had to involve the same driver each day. And gas or travel expense could be reimbursed no more frequently than weekly.

Trentway-Wagar was arguing that because carpooling used to be inconvenient, it should always be inconvenient, and if that inconvenience disappeared, then it should be reinserted by legal fiat. Curiously, an organization that commits to helping society manage a problem also commits itself to the preservation of that same problem, as its institutional existence hinges on society's continued need for its management. Bus companies provide a critical service—public transportation—but they also commit themselves, as Trentway-Wagar did, to fending off competition from alternative ways of moving people from one place to another.

The OHTB upheld Trentway-Wagar's complaint and ordered PickupPal to stop operating in Ontario. PickupPal decided to fight the case—and lost in the hearing. But public attention became focused on the issue, and in a year of high gas prices, burgeoning environmental concern, and a financial downturn, almost no one took Trentway-Wagar's side. The public reaction, channeled through everything from an online petition to T-shirt sales, had one message: Save PickupPal. The idea that people couldn't use such a service was too hot for the politicians in Ontario to ignore. Within weeks of Trentway-Wagar's victory, the Ontario legislature amended the Public Vehicles Act to make PickupPal legal again.

PickupPal makes use of social media in several ways. First and foremost, it provides its users with enough information quickly enough that they can coordinate to solve a real-world problem. PickupPal simply could not exist in the absence of a medium that allowed potential drivers and riders to share information about their respective routes. Second, it creates aggregate value—the more numerous its users, the greater the likelihood of a match.

Old logic, television logic, treated audiences as little more than collections of individuals. Their members didn't create any real value for one another. The logic of digital media, on the other hand, allows the people formerly known as the audience to create value for one another every day.

PickupPal also relies on the erasure of the old distinction between online media and "the real world." It is an online service in only the most trivial way—it produces value for its users by matching them up; but that value is realized only when an actual rider and an actual driver share an actual car on an actual highway. This is a case of social media as part of the real world, as a way of improving the real world, in fact, rather than standing apart from it. The use of publicly available media as a coordinating resource for thousands of ordinary citizens marks a departure from the media landscape we're used to. The public media we're most familiar with, of course, is the twentieth-century model, with professional producers and amateur consumers. Its underlying economic and institutional logic started not in the twentieth century, but in the fifteenth.

GUTENBERG ECONOMICS

Johannes Gutenberg, a printer in Mainz, in present-day Germany, introduced movable type to the world in the middle of the fifteenth century. Printing presses were already in use, but they were slow and laborious to operate, because a carving had to be made of the full text of each page. Gutenberg realized that if you made carvings of individual letters instead, you could arrange them into any words you liked. These carved letters—type—could be moved around to make new pages, and the type could be set in a fraction of the time that it would take to carve an entire page from scratch.

Movable type introduced something else to the intellectual landscape of Europe: an abundance of books. Prior to Gutenberg, there just weren't that many books. A single scribe, working alone with a quill and ink and a pile of vellum, could make a copy of a book, but the process was agonizingly slow, making output of scribal copying small and the price high. At the end of the fifteenth century, a scribe could produce a single copy of a five-hundred-page book for roughly thirty florins, while Ripoli, a Venetian press, would, for roughly the same price, print more than three hundred copies of the same book. Hence most scribal capacity was given over to producing additional copies of extant works. In the thirteenth century Saint Bonaventure, a Franciscan monk, described four ways a person could make books: copy a work whole, copy from several works at once, copy an existing work with his own additions, or write out some of his own work with additions from elsewhere. Each of these categories had its own name, like scribe or author, but Bonaventure does not seem to have considered—and certainly didn't describe—the possibility of anyone creating a wholly original work. In this period, very few books were in existence and a good number of them were copies of the Bible, so the idea of bookmaking was centered on re-creating and recombining existing words far more than on producing novel ones.

Movable type removed that bottleneck, and the first thing the growing cadre of European printers did was to print more Bibles—lots more Bibles. Printers began publishing Bibles translated into vulgar languages—contemporary languages other than Latin—because priests wanted them, not just as a convenience but as a matter of doctrine. Then they began putting out new editions of works by Aristotle, Galen, Virgil, and others that had survived from antiquity. And still the presses could produce more. The next move by the printers was at once simple and astonishing: print lots of new stuff. Prior to movable type, much of the literature available in Europe had been in Latin and was at least a millennium old. And then in a historical eyeblink, books started appearing in local languages, books whose text was months rather than centuries old, books that were, in aggregate, diverse, contemporary, and vulgar. (Indeed, the word *novel* comes from this period, when newness of content was itself new.)

This radical solution to spare capacity—produce books that no one had ever read before—created new problems, chiefly financial risk. If a printer produced copies of a new book and no one wanted to read it, he'd lose the resources that went into creating it. If he did that enough times, he'd be out of business. Printers reproducing Bibles or the works of Aristotle never had to worry that people might not want their wares, but anyone who wanted to produce a novel book faced this risk. How did printers manage that risk?

Their answer was to make the people who bore the risk—the printers—responsible for the quality of the books as well. There's no obvious reason why people who are good at running a printing press should also be good at deciding which books are worth printing. But a printing press is expensive, requiring a professional staff to keep it running, and because the material has to be produced in advance of demand for it, the economics of the printing press put the risk at the site of production. Indeed, shouldering the possibility that a book might be unpopular marks the transition from printers (who made copies of hallowed works) to publishers (who took on the risk of novelty).

A lot of new kinds of media have emerged since Gutenberg: images and sounds were encoded onto objects, from photographic plates to music CDs; electromagnetic waves were harnessed to create radio and TV. All these subsequent revolutions, as different as they were, still had the core of Gutenberg economics: enormous investment costs. It's expensive to own the means of production, whether it is a printing press or a TV tower, which makes novelty a fundamentally high-risk operation. If it's expensive to own and manage the means of production or if it requires a staff, you're in a world of Gutenberg economics. And wherever you have Gutenberg economics, whether you are a Venetian publisher or a Hollywood producer, you're going to have fifteenth-century risk management as well, where the producers have to decide what's good before showing it to the audience. In this world almost all media was produced by "the media," a world we all lived in up until a few years ago.

THE BUTTON MARKED "PUBLISH"

At the end of every year, the National Book Foundation hands out its medal for Distinguished Contribution to American Letters at its awards dinner. In 2008 it gave the award to Maxine Hong Kingston, author of 1976's *The Woman Warrior*. While Kingston was being recognized for work that was more than thirty years old, her speech included a retelling of something she'd done that year, something that should have made the blood of every publisher in attendance run cold.

Earlier that year, Kingston said, she had written an editorial praising Barack Obama, on the occasion of his visit to her home state of Hawaii. Unfortunately for her, the newspapers she sent the piece to all declined to publish it. And then, to her delight, she realized that this rejection mattered a whole lot less than it used to. She went onto Open.Salon.com, a website for literary conversation, and, as she put it, "All I had to do was type, then click a button marked 'Publish.' Yes, there is such a button. Voilà! I was published."

Yes, there is such a button. Publishing used to be something we had to ask permission to do; the people whose permission we had to ask were publishers. Not anymore. Publishers still perform other functions in selecting, editing, and marketing work (dozens of people besides me have worked to improve this book, for example), but they no longer form the barrier between private and public writing. In Kingston's delight at routing around rejection, we see a truth, always there but long hidden. Even "published authors," as the phrase goes, didn't control their own ability to publish. Consider the cluster of ideas contained in this list: publicity, publicize, publish, publication, publicist, publisher. They are all centered on the act of making something public, which has historically been difficult, complex, and expensive. And now it is none of those things.

Kingston's editorial, it must be said, wasn't any good. It was obsequious to the point of tedium and free of any thought that might be called analytic. The political discourse was not much enriched by its appearance. But an increase in freedom to publish always has this consequence. Before Gutenberg, the average book was a masterpiece. After Gutenberg, people got throwaway erotic novels, dull travelogues, and hagiographies of the landed gentry, of interest to no one today but a handful of historians. The great tension in media has always been that freedom and quality are conflicting goals. There have always been people willing to argue that an increase in freedom to publish isn't worth the decrease in average quality; Martin Luther observed in 1569: "The multitude of books is a great evil. There is no measure of limit to this fever for writing; every one must be an author; some out of vanity, to acquire celebrity and raise up a name; others for the sake of mere gain." Edgar Allan Poe commented in 1845: "The enormous multiplication of books in every branch of knowledge is one of the greatest evils of this age; since it presents one of the most serious obstacles to the acquisition of correct information by throwing in the reader's way piles of lumber in which he must painfully grope for the scraps of useful lumber."

These arguments are absolutely correct. Increasing freedom to publish does diminish average quality—how could it not? Luther and Poe both relied on the printing press, but they wanted the mechanics of publishing, to which they had easy access, not to increase the overall volume of published work: cheaper for me but still inaccessible to thee. Economics doesn't work that way, however. The easier it is for the average person to publish, the more average what gets published becomes. But increasing freedom to participate in the public conversation has compensating values.

The first advantage is an increase of experimentation in form. Even though the spread of movable type created a massive downshift in average quality, that same invention made it possible to have novels, newspapers, and scientific journals. The press allowed the rapid dissemination of both Martin Luther's *Ninety-five Theses* and Copernicus's *On the Revolutions of the Celestial Spheres*, transformative documents that influenced the rise of the Europe we know today. Lowered costs in any realm allow for increased experimentation; lowered costs for communication mean new experimentation in what gets thought and said.

This ability to experiment extends to creators as well, increasing not just their number but also their diversity. Naomi Wolf, in her 1991 book *The Beauty Myth,* both celebrated and lamented the role women's magazines play in women's lives. These magazines, she said, provide a place where a female perspective can be taken for granted, but it is distorted by the advertisers: "Advertisers are the West's courteous censors. They blur the line between editorial freedom and the demands of the marketplace . . . A women's magazine's profit does not come from its cover price, so its contents cannot roam too far from the advertiser's wares." Today, on the other hand, almost twenty years after *The Beauty Myth* appeared, writer Melissa McEwan posted on the blog Shakesville a riveting seventeen-hundred-word essay about casual misogyny:

> There are the jokes about women . . . told in my presence by men who are meant to care about me, just to get a rise out of me, as though I am meant to find funny a reminder of my second-class status. I am meant to ignore that this is a bullying tactic, that the men telling these jokes derive their amusement specifically from knowing they upset me, piss me off, hurt me. They tell them and I can laugh, and they can thus feel superior, or I can not laugh, and they can thus feel superior. Heads they win, tails I lose.

The essay, titled "The Terrible Bargain We Have Regretfully Struck," attracted hundreds of commenters and thousands of readers in an outpouring of reaction whose main theme was *Thank you for saying what I have been thinking.* The essay got out into the world because McEwan only had to click a button marked "Publish." Shakesville provides exactly the kind of writing space Wolf imagined, where women can talk without male oversight or advertisers' courteous censorship. The writing is not for everyone—intensely political, guaranteed to anger any number of people—but that's exactly the point. The women's magazines Wolf discussed reached readers who might have had the same reaction as the readers of Shakesville, but the magazines simply couldn't afford to reach them at the expense of angering other readers or, more important, their advertisers. McEwan was willing (and able) to risk angering people in order to say what she had to say.

The bargain Wolf described was particularly acute for women's magazines, but it was by no means unique. Nor is the self-publishing model McEwan used unique—people now speak out on issues a million times a day, across countless kinds of communities of interest. The ability for community members to speak to one another, out loud and in public, is a huge shift, and one that has value even in the absence of a way to filter for quality. It has value, indeed, *because* there is no way to filter for quality in advance: the definition of quality becomes more variable, from one community to the next, than when there was broad consensus about mainstream writing (and music, and film, and so on).

Scarcity is easier to deal with than abundance, because when something becomes rare, we simply think it more valuable than it was before, a conceptually easy change. Abundance is different: its advent means we can start treating previously valuable things as if they were cheap enough to waste, which is to say cheap enough to experiment with. Because abundance can remove the trade-offs we're used to, it can be disorienting to the people who've grown up with scarcity. When a resource is scarce, the people who manage it often regard it as valuable in itself, without stopping to consider how much of the value is tied to its scarcity. For years after the price of long-distance phone calls collapsed in the United States, my older relatives would still announce that a call was "long distance." Such calls had previously been special, because they were expensive; it took people years to understand that cheap long-distance calls removed the rationale for regarding them as inherently valuable.

Similarly, when publication—the act of making something public—goes from being hard to being virtually effortless, people used to the old system often regard publishing by amateurs as frivolous, as if publishing was an inherently serious activity. It never was, though. Publishing had to be taken seriously when its cost and effort made people take it seriously—if you made too many mistakes, you were out of business. But if these factors collapse, then the risk collapses too. An activity that once seemed inherently valuable turned out to be only accidentally valuable, as a change in the economics revealed.

Harvey Swados, the American novelist, said of paperbacks, "Whether this revolution in the reading habits of the American public means that we are being inundated by a flood of trash which will debase farther the popular taste, or that we shall now have available cheap editions of an ever-increasing list of classics, is a question of basic importance to our social and cultural development."

He made this observation in 1951, two decades into the spread of paperbacks, and curiously Swados was even then unable to answer his own question. But by 1951 the answer was plain to see. The public had no need to choose between a flood of trash and a growing collection of classics. We could have both (which is what we got).

Not only was "both" the answer to Swados's question; it has always been the answer whenever communications abundance increases, from the printing press on. The printing press was originally used to provide cheap access to Bibles and the writings of Ptolemy, but the entire universe of that old stuff didn't fill a fraction of either the technological capacity or the audience's desire. Even more relevant to today, we can't have "an ever-expanding list of classics" without also trying new forms; if there was an easy formula for writing something that will become prized for decades or centuries, we wouldn't need experimentation, but there isn't, so we do.

The low-quality material that comes with increased freedom accompanies the experimentation that creates the stuff we will end up prizing. That was true of the printing press in the fifteenth century, and it's true of the social media today. In comparison with a previous age's scarcity, abundance brings a rapid fall in average quality, but over time experimentation pays off, diversity expands the range of the possible, and the best work becomes better than what went before. After the printing press, publishing came to matter more because the expansion of literary, cultural, and scientific writing benefited society, even though it was accompanied by a whole lot of junk.

THE CONNECTIVE TISSUE OF SOCIETY

Not that we are witnessing a rerun of the print revolution. All revolutions are different (which is only to say that all surprises are surprising). If a change in society were immediately easy to understand, it wouldn't be a revolution. And today, the revolution is centered on the shock of the inclusion of amateurs as producers, where we no longer need to ask for help or permission from professionals to say things in public. Social media didn't cause the candlelight protests in South Korea; nor did they make users of PickupPal more environmentally conscious. Those effects were created by citizens who wanted to change the way public conversation unfolded and found they had the opportunity to do so.

This ability to speak publicly and to pool our capabilities is so different from what we're used to that we have to rethink the basic concept of media: it's not just something we consume, it's something we use. As a result, many of our previously stable concepts about media are now coming unglued.

Take, as one example, television. Television encodes moving images and sounds for transmission through the air and, latterly, through a cable, for subsequent conversion back to images and sound, using a special decoding device. What is the name of the content so transmitted? Television. And the device that displays the images? It is a television. And the people who make that content and send out the resulting signal—what industry do they work in? Television, of course. The people who work in television make television for your television.

You can buy a television at the store so you can watch television at home, but the television you buy isn't the television you watch, and the television you watch isn't the television you buy. Expressed that way, it seems confusing, but in daily life it isn't confusing at all, because we never have to think too hard about what television is, and we use the word *television* to talk about all the various different parts of the bundle: industry, content, and appliance. Language lets us work at the right level of ambiguity; if we had to think about every detail of every system in our lives all the time, we'd faint from overexposure. This bundling of object and industry, of product and service and business model, isn't unique to television. People who collect and preserve rare first editions of books, and people who buy mass-market romance novels, wreck the spines, and give them away the next week, can all legitimately lay claim to the label book lover.

This bundling has been easy because so much of the public media environment has been stable for so long. The last really big revolution in public media was the appearance of television. In the sixty years since TV went mainstream, the kinds of changes we've seen have been quite small—the spread of videocassette tapes, for example, or color TV. Cable television was the most significant change in the media landscape between the late 1940s (when TV started to spread in earnest) and the late 1990s (when digital networks began to be a normal part of public life).

The word *media* itself is a bundle, referring at once to process, product, and output. Media, as we talked about it during those decades, mainly denoted the output of a set of industries, run by a particular professional class and centered, in the English-speaking world, in London, New York, and Los Angeles. The word referred to those industries, to the products they created, and to the effect of those products on society. Referring to "the media" in that way made sense as long as the media environment was relatively stable.

Sometimes, though, we really do have to think about the parts of a system separately, because the various pieces stop working together. If you take five minutes to remind yourself (or conjure up, if you are under thirty) what media for adults was like in the twentieth century, with a handful of TV networks and dominant newspapers and magazines, then media today looks strange and new. In an environment so stable that getting TV over a wire instead of via antennae counted as an upheaval, it's a real shock to see the appearance of a medium that lets anyone in the world make an unlimited number of perfect copies of something they created for free. Equally surprising is the fact that the medium mixes broadcast and conversational patterns so thoroughly that there is no obvious gulf between them. The bundle of concepts tied to the word *media* is unraveling. We need a new conception for the word, one that dispenses with the connotations of "something produced by professionals for consumption by amateurs."

Here's mine: media is the connective tissue of society.

Media is how you know when and where your friend's birthday party is. Media is how you know what's happening in Tehran, who's in charge in Tegucigalpa, or the price of tea in China. Media is how you know what your colleague named her baby. Media is how you know why Kierkegaard disagreed with Hegel. Media is how you know where your next meeting is. Media is how you know about anything more than ten yards away. All these things used to be separated into public media (like visual or print communications made by a small group of professionals) and personal media (like letters and phone calls made by ordinary citizens). Now those two modes have fused.

The internet is the first public medium to have post-Gutenberg economics. You don't need to understand anything about its plumbing to appreciate how different it is from any form of media in the previous five hundred years. Since all the data is digital (expressed as numbers), there is no such thing as a copy anymore. Every piece of data, whether an e-mailed love letter or a boring corporate presentation, is identical to every other version of the same piece of data.

You can see this reflected in common parlance. No one ever says, *Give me a copy of your phone number.* Your phone number is the same number for everybody, and since data is made of numbers, the data is the same for everybody. Because of this curious property of numbers, the old distinction between copying tools for professionals and those for amateurs—printing presses that make high-quality versions for the pros, copy machines for the rest of us—is over. Everyone has access to a medium that makes versions so identical that the old distinction between originals and copies has given way to an unlimited number of equally perfect versions.

Moreover, the means of digital production are symmetrical. A television station is a hugely expensive and complex site designed to send signals, while a television is a relatively simple device for receiving those signals. When someone buys a TV, the number of consumers goes up by one, but the number of producers stays the same. On the other hand, when someone buys a computer or a mobile phone, the number of consumers and producers both increase by one. Talent remains unequally distributed, but the raw ability to make and to share is now widely distributed and getting wider every year.

Digital networks are increasing the fluidity of all media. The old choice between one-way public media (like books and movies) and two-way private media (like the phone) has now expanded to include a third option: two-way media that operates on a scale from private to public. Conversations among groups can now be carried out in the same media environments as broadcasts. This new option bridges the two older options of broadcast and communications media. All media can now slide from one to the other. A book can stimulate public discussion in a thousand places at once. An e-mail conversation can be published by its participants. An essay intended for public consumption can anchor a private argument, parts of which later become public. We move from public to private and back again in ways that weren't possible in an era when public and private media, like the radio and the telephone, used different devices and different networks.

And finally, the new media involves a change in economics. With the internet, everyone pays for it, and then everyone gets to use it. Instead of having one company own and operate the whole system, the internet is just a set of agreements about how to move data between two points. Anyone who abides by these agreements, from an individual working from a mobile phone to a huge company, can be a full-fledged member of the network. The infrastructure isn't owned by the producers of the content: it's accessible to everyone who pays to use the network, regardless of how they use it. This shift to post-Gutenberg economics, with its interchangeably perfect versions and conversational capabilities, with its symmetrical production and low costs, provides the means for much of the generous, social, and creative behavior we're seeing.

THREE AMATEURS WALK INTO A BAR

Because all the public media we've known until recently abided by Gutenberg economics, we assumed, without even really thinking about it, that media needed professionals to guarantee its very existence. We assumed that we audience members weren't just relegated to consuming but preferred that status. With this implicit theory of the media landscape in our heads, generous, public, and creative behavior does indeed look puzzling, at the very least. Like so many surprising behaviors, this one comes mainly from mistaking an accidental pattern for a deep truth.

People sharing their writing or their videos or their medical symptoms or seats in their car are motivated by something other than the desire for money. The people running services like YouTube and Facebook want to get paid, and they do. It can seem unfair for amateurs to be contributing their work for free to people who are making money from aggregating and sharing that work. At least traditional media outlets pay their contributors; with the new services, which enable amateurs to share work, the revenue goes not to the content creators but to the owners of the platform that enables the sharing, leading to the obvious question: why are all these people working for free? The writer Nicholas Carr has dubbed this pattern digital sharecropping, after the post–Civil War sharecroppers who worked the land but didn't own it or the food they grew on it. With digital sharecropping, the platform owners get the money and the creators of the content don't, a situation Carr regards as manifestly unfair.

Curiously, the people most affected by this state of affairs don't seem to be terribly up in arms about it. The people sharing photos and videos and writing don't expect to be paid, but they share anyway. The complaints about digital sharecropping arise partly from professional jealousy—clearly professional media makers are upset about competition from amateurs. But there's another, deeper explanation: we're using a concept from professional media to refer to amateur behaviors, but amateurs' motivations differ from those of professionals. If ICanHasCheezburger.com, purveyor of lolcats, is a late-model version of the fifteenth-century publishing model, then the fact that its workers are contributing their labor unpaid is not just strange but unfair. But what if the contributors aren't workers? What if they really are contributors, quite specifically intending their contributions to be acts of sharing rather than production? What if their labors are labors of love?

ICanHasCheezburger may look like a traditional media outlet, but that doesn't mean it has the same internal logic as a professional outlet like *Time* magazine. Consider, as an alternative comparison, a local bar. It's a commercial operation, but the products it sells are invariably cheaper at home, often by a considerable margin; much of the service offered by the staff amounts to opening bottles and washing dishes. If a beer costs twice as much from a bar as it does from a store, why doesn't the whole business just collapse as people opt for cheaper beer at home?

Like the owners of YouTube, the bar owner is in the curious business of offering value above the products and services he sells, value that is created by the customers for one another. People pay more to have a beer in a bar than they do at home because a bar is a more convivial place to have a drink; it draws in people who are seeking a little conversation or just want to be around other people, people who prefer being in the bar to being home alone. This inducement is powerful enough that the difference is worth paying for. The digital sharecropper logic would suggest that the bar owner is exploiting his customers, because their conversations in the bar are part of the "content" that makes them willing to overpay for the beer, but none of the customers actually feels that way. Instead, they willingly reward the owner for creating a socially welcoming environment, a place where they will pay extra for the chance to associate with one another.

However, the digital sharecropping logic does sometimes apply; people can sometimes feel precisely as Carr predicts they should. One of the largest instances of digital sharecropping backlash came from volunteers for the online service America Online. In the 1980s and 1990s AOL grew largely because people found its friendly and helpful image appealing. Its Community Leaders, an all-volunteer corps, were always present in public chat rooms and other areas of the site, guiding discussions, watching out for insulting or offensive language, and generally keeping things on an even keel. In 1999 two of these guides, Brian Williams and Kelly Hallisey, sued AOL on behalf of the ten thousand or so other volunteers, arguing that they should have been paid at least minimum wage for their labor.

Given that Williams, Hallisey, and indeed all the Leaders had willingly volunteered their time, and had done so for years (Williams estimated his time ran to several thousand hours), it's hard to see why they would later decide they'd been abused. The answer, as in any relationship that goes sour, lies in what changed. In an interview, Williams blamed the commercialization of the service. "It increasingly became like AOL was just trying to make a dollar off the back of free slave labor. Before, you didn't have advertising everywhere, and it was a much richer community where people got together to get together, and now it's not like that." The change from a community-driven site to an advertising-driven site changed the feelings of the Leaders; they started applying the digital sharecropping logic only when AOL stopped providing visible appreciation. (The suit, now in its second decade, includes thousands of former Community Leaders and has not yet been settled.)

Humans intrinsically value a sense of connectedness; given that reality, the digital sharecropping logic loses much of its explanatory power. Amateurs aren't just pint-sized professionals;

people are sometimes happy to do things for reasons that are incompatible with getting paid. Amateur media is different from professional media. Services that help us share things thrive precisely because they make it easier and often cheaper for us to do things we're already inclined to do. One function of the market, in other words, is to provide platforms for us to engage in the things we value doing outside the market, whether those platforms are bars or websites. The fifteenth-century model of media production didn't allow for that kind of sharing, because its inherent cost and risk meant professionals were required at every step. Now they're not.

THE SHOCK OF INCLUSION

I teach in the Interactive Telecommunications Program, an interdisciplinary graduate program at NYU. In the decade I've been there, the average age of my students has stayed roughly the same, while my average age has grown at the alarming rate of one year per year; my students are now fifteen or twenty years younger than I am. Because I try to convey an understanding of the changing media landscape, I now have to teach the times of my own youth as ancient history. Seemingly stable parts of the world I grew up in had vanished before many of my students turned fifteen, while innovations I saw take hold with adult eyes occurred when they were in grade school.

Despite half a century of hand-wringing about media contraction, my students have never known a media landscape of anything less than increasing abundance. They have never known a world with only three television channels, a world where the only choice a viewer had in the early evening was which white man was going to read them the news in English. They can understand the shift from scarcity to abundance, since the process is still going on today. A much harder thing to explain to them is this: if you were a citizen of that world, and you had something you needed to say in public, you couldn't. Period. Media content wasn't produced by consumers; if you had the wherewithal to say something in public, you weren't a consumer anymore, by definition. Movie reviews came from movie reviewers. Public opinions came from opinion columnists. Reporting came from reporters. The conversational space available to mere mortals consisted of the kitchen table, the water cooler, and occasionally letter writing (an act so laborious and rare that many a letter began with "Sorry I haven't written in so long . . .")

In those days, anyone could produce a photograph, a piece of writing, or a song, but they had no way to make it widely available. Sending messages to the public wasn't for the public to do, and, lacking the ability to easily connect with one another, our motivation to create was subdued. So restricted was access to broadcast and print media that amateurs who tried to produce it were regarded with suspicion or pity. Self-published authors were assumed to be either rich or vain. People who published pamphlets or walked around with signs were assumed to be unhinged. William Safire, the late columnist for *The New York Times*, summed up this division: "For years I used to drive up Massachusetts Avenue past the vice president's house and would notice a lonely, determined guy across the street holding a sign claiming he'd been sodomized by a priest. Must be a nut, I figured—and thereby ignored a clue to the biggest religious scandal of the century."

My students believe me when I tell them about the assumed

silence of the average citizen. But while they are perfectly able to make intellectual sense of that world, I can tell they don't feel it. They've never lived in an environment where they weren't able to speak in public, and it's hard for them to imagine how different that environment was, compared with the participatory behaviors they take for granted today.

Nik Gowing, a BBC reporter and author of *"Skyful of Lies" and Black Swans,* about media in crises, offers an illustrative story. In the hours after the London subway and bus bombings of July 7, 2005, the government maintained that the horrific damage and casualties had been caused by some sort of power surge. Even a few years earlier, this explanation would have been the only message available to the public, allowing the government time to investigate the incident more fully before adjusting its story to reflect the truth. But as Gowing notes, "Within the first 80 minutes in the public domain, there were already 1,300 blog posts signaling that explosives were the cause."

The government simply could not stick to the story about a power surge when its falsehood was increasingly apparent to all. Camera-phones and sites for sharing photos globally meant that the public could see images of the subway interior and of a double-decker bus whose roof had been blown to pieces—evidence utterly incompatible with the official story. Less than two hours after the bombings, Sir Ian Blair, the Metropolitan Police commissioner, publicly acknowledged that the explosions had been the work of terrorists. He did so even though his grasp of the situation wasn't yet complete, and against the advice of his aides, simply because people were already trying to understand the events without waiting for him to speak. The choice for the police had previously been

Should we tell the public something or nothing? By 2005, it had become Do we want to be part of the conversation the public is already having? Blair decided to speak to the public at that early stage because the older strategies that assumed that the public wasn't already talking among itself were no longer intact.

The people surprised at our new behaviors assume that behavior is a stable category, but it isn't. Human motivations change little over the years, but opportunity can change a little or a lot, depending on the social environment. In a world where opportunity changes little, behavior will change little, but when opportunity changes a lot, behavior will as well, so long as the opportunities appeal to real human motivations.

The harnessing of our cognitive surplus allows people to behave in increasingly generous, public, and social ways, relative to their old status as consumers and couch potatoes. The raw material of this change is the free time available to us, time we can commit to projects that range from the amusing to the culturally transformative. If free time was all that was necessary, however, the current changes would have occurred half a century ago. Now we have the tools at our disposal, and the new opportunities they provide.

Our new tools haven't caused those behaviors; but they have allowed them. Flexible, cheap, and inclusive media now offers us opportunities to do all sorts of things we once didn't do. In the world of "the media," we were like children, sitting quietly at the edge of a circle and consuming whatever the grown-ups in the center of the circle produced. That has given way to a world in which most forms of communication, public and private, are available to everyone in some form. Even accepting that these new behaviors are happening and that new kinds of media are providing the means for them, we still have to explain why. New tools get used only if they help people do things they want to do; what is motivating The People Formerly Known as the Audience to start participating?

3

Motive

Josh Groban is a classically trained American baritone who sings what's sometimes called classical crossover or popera, a style of music that sounds pretty much like what you'd expect—soulful renditions of pop songs in Italian and English ("Alla Luce del Sole," "Per Te," "You Raise Me Up") with a few operatic standards like "Ave Maria" thrown in. He is terrifically successful; all four of his U.S. albums to date have sold two million copies or more. Groban is talented, emotionally demonstrative, and cute; his legions of fans have been described as teenage girls and their grandmothers. He has the sort of audience, in other words, that couldn't have been easily assembled using traditional media, because there are no radio stations catering to that range of ages.

That makes Groban a good old internet success story. As with Dong Bang Shin Ki, his existing fans often recruit new ones, and word-of-mouth marketing unfolds in a media landscape increasingly created by the fans themselves. You can see their engagement on JoshGroban.com, where a group of hard-core fans refer to themselves as Grobanites and to all things Josh as Grobania.

The story of an artist using the web to find fans is familiar by now; what's interesting is something that happened once those fans got together. In 2002 a few Grobanites proposed getting Groban something for his twenty-first birthday. The choice of gift presented the fans with a dilemma: the object of their affection, after all, was a young man who, before he was legally allowed to drink beer, had already achieved fame, fortune, and endless adulation. What could he possibly want? After the Grobanites discussed and rejected several ideas (how many teddy bears does one man need?), one Grobanite, Julie Clarke, suggested passing the hat and making a charitable donation in his name. They decided that whatever funds they raised should go to the David Foster Foundation, a charity run by Groban's producer that works with disadvantaged youth. Clarke agreed to handle the donations, eventually collecting over a thousand dollars. Groban was happily surprised, the David Foster Foundation was delighted, and the people who donated felt the warm glow of accomplishment.

Seeing this success, Clarke and another Grobanite, Valerie Sooky, whom she had met during the fund-raising drive, worked to make charitable giving part of life in Grobania. Whenever Groban is on tour, some of his fans gather before the concert for a meet-and-greet. Grobanites started collecting donations at these gatherings, often raising hundreds of dollars at a time. These events provide a place for fans to gather once a year or so, but JoshGroban.com does that every day. So Clarke suggested hosting an online charity auction for Groban's next birthday. She and Sooky recruited Megan Markus, a nineteen-year-old Grobanite eager to help, to design an auction site. It was a distinctly amateur affair: it was Markus's first web design job, many of the goods were made and donated by the Grobanites themselves, and the system was so cumbersome, all bids had to be entered into the system by hand. Finally, after weeks of learning how to run an unfamiliar piece of software, they launched their first auction.

They raised $16,000 in a few days—an order of magnitude more than they'd raised for any other event. Then they had another auction. And another. Over the course of a year, they raised $75,000, culminating in Groban's next birthday donation, for which they raised nearly $24,000 in a single event.

Clarke, Sooky, and Markus realized they were on to something. The money was still coming in, and while everyone was pleased to support the David Foster Foundation, the funds weren't being raised by Fosterites, so they asked Groban how they could work together more closely. This was a challenge for Groban's lawyers, if only because of the novelty—charitable organizations set up by entertainers are typically funded by the star, so there's no clear precedent for taking donations from fans in any sort of organized way.

Eventually Groban's lawyers created a 501(c)3, a nonprofit corporation, with the serviceable if unimaginative name The Josh Groban Foundation. Its principal function was to act as a kind of legitimate "money-laundering" service, allowing the charitably minded Grobanites to make donations in Groban's name, with the foundation receiving and then disbursing the results. This arrangement worked well enough for a while—the Grobanites continued to raise money and identify worthy new recipients (now in collaboration with the foundation).

By 2004 this group of generous Grobanites was growing faster than the foundation itself, which, in keeping with its function as an engine of disbursement, offered no organizational face that people could connect with, not even an e-mail address. As the Grobanites grew in numbers, the internal management issues became more complex. (This always happens when groups grow in size, age, or ambition; with the Grobanites, all three things happened at once.) The founders talked about how to handle this newfound complexity: should they become the volunteer wing of the Josh Groban Foundation, or start their own organization? After over a year of discussion—ad hoc groups usually involve a lot of talking—the fact that they were fans reaching out to other fans tipped their decision ("We are one of them, they know us, they trust us," as Sooky later described the reasoning to me), and they formed their own nonprofit organization, Grobanites for Charity.

The original birthday fund-raiser, and the social commitment it kicked off, eventually drove the creation of not one but two organizations that now function as two halves of a single whole—Grobanites for Charity raised the funds, and the Josh Groban Foundation disbursed them. Compared with traditional charities, the Grobanites did everything backward. The usual nonprofit model, for groups like Save the Children and the Sierra Club, assumes the organization is formed first, and then acquires members; the institution is in place before it begins to raise money. Grobanites for Charity had members before it had a mission, its members raised money before they had an institution, and the founders created an institution even after someone else had settled all the legal issues.

Further, their success begat success—other groups of Grobanites started pursuing charitable work as well. Grobanites for Africa, a wholly unowned subsidiary of Grobanites for Charity, is specifically dedicated to raising money for organizations fighting poverty and the effects of HIV/AIDS on that continent. This group started after Groban's first international concert tour took him to South Africa, where he met Nelson Mandela and announced his support for charitable work on behalf of African children. A group of Grobanites, preparing the meet-and-greets for a tour stop in Atlanta, decided to adopt this cause and, true to form, organized themselves separately; they work closely with other Grobanites and with the Josh Groban Foundation, a pattern established by the original fund-raising efforts. To date, Grobanites for Africa have generated over $150,000 for those causes.

The important question about Grobanites for Charity isn't, Where did they find the time to form a charity? We know the Grobanites had free time and access to media that connected them when they wanted to be connected. Nor is the question, How did they come to be part of a coordinated group? That answer too is familiar: JoshGroban.com created a place where people could come together, share their ideas and goals, and egg each other on. Once they'd started working together, it also gave them an environment to solicit like-minded Grobanites.

The puzzling question is Why? Why did these women set themselves the task of raising money in the first place, and why would the Grobanites for Charity create a separate entity for themselves even though the Josh Groban Foundation already existed? This isn't lolcats; running Grobanites for Charity is hard work, and not only are the participants unpaid, they're putting their own money into the effort. Of all the things to do online, what would motivate someone to give up this much of her own time and money for something that produces no obvious tangible reward?

LOVE OVER GOLD

In 1970 Edward Deci, a research psychologist at the University of Rochester, performed a remarkably simple experiment that ignited a controversy still roiling today. The experiment was based on a puzzle game called Soma, a wooden cube subdivided into seven smaller pieces. Each of the seven pieces is unique; there's a T-shape, an L-shape, and so on. These seven pieces can be assembled into the larger cube in only one way; they can also be put together to make millions of other shapes. The challenge with a Soma cube is to look at a drawing of a potential shape, then figure out how to assemble the seven pieces to create that shape. It's harder than it sounds. Deci built his observations on this challenge.

At the beginning of the experiment, Deci gave a subject the Soma pieces and diagrams of three or four of the shapes that could be assembled. Once the subject familiarized himself with the pieces (the subjects were all men), Deci asked him to assemble the pieces into the three or four shapes in the diagrams but did not say how. Deci repeated this process with dozens of subjects, who all thought assembling the shapes was the experiment. It wasn't. After giving the explanation and observing the student for most of the hour, Deci would leave the room, telling the subject to take a break and wait there until he returned. During his absence from the room Deci observed the subject through a one-way mirror for exactly eight minutes. The subject's behavior during that break was the experiment.

While Deci was gone, the subjects were free to choose their own activities. Deci had placed magazines and other distractions in the experiment room. (Since this was 1970, the distractions included issues of *The New Yorker, Time,* and *Playboy* and an ashtray.) Even with these items readily available, many of the students kept playing with the puzzle on their own, spending on average about half of the eight minutes working on it. When Deci returned, he would release the students, having recorded their break-time activities, behavior that provided Deci with a baseline for the participants' voluntary engagement with the puzzle.

Deci then invited the same subjects to a second session of Soma challenges, with one difference. This time he asked half of them to work with the puzzle exactly as before. The other dozen, though, were told that they would be paid a dollar for every shape they assembled (in an era when a dollar was worth something to a college student). Again they were told to take a break, during which they were secretly observed for eight minutes while alone in the room. The paid subjects, who now thought of the cubes as a potential source of income, experimented with them, on average, for a minute more of their break time than they had previously. Deci then ran a third session, where he simply repeated the experiment exactly as he had run it initially: all the subjects were asked to assemble shapes, with no pay for anyone. In this session, even though each subject received identical instructions, the ones who had been paid in the previous session showed markedly less interest in the shapes during the break than in the session where they had been paid; their average time spent dropped by two minutes, which is to say it fell twice as far, when they payment was removed, as it had risen when the payment was added in the first place. Even though they had played with the puzzle voluntarily in the first session, the memory of having been paid earlier was enough to depress their interest when they were again given the chance to experiment with the puzzle on their own.

In psychological literature, experiments designed to illuminate voluntary engagement are called "free choice" tests—when someone has control over his actions, how likely is he to engage in a particular behavior? Deci's Soma experiment found that payment for working with the puzzle depressed free choice for the same activity. Deci's conclusion was that human motivation isn't purely additive. Doing something because it interests you makes it a different kind of activity than doing it because you are reaping an external reward. The experiment substantiated a psychological theory that distinguished between two broad types of motivation, intrinsic and extrinsic. Intrinsic motivations are those in which the activity itself is the reward. In the case of Soma, the subjects who kept working on the puzzle during their break were clearly motivated by the satisfaction that would come from doing so correctly. Extrinsic motivations are those in which the reward for doing something is external to the activity, not the activity itself. Payment is the classic case of extrinsic motivation, which is why the subjects were paid to assemble the shapes.

Receiving sufficient payment can make otherwise undesirable activity desirable and worthwhile. (Thus is society able to employ garbage collectors.) But Deci's experiment suggested that extrinsic motivations aren't always the most effective ones and that increasing extrinsic motivations can actually decrease intrinsic ones. He concluded that an extrinsic motivation like being paid can crowd out an intrinsic one like enjoying something for its own sake. (This idea of one motivation's crowding out another also appears in the literature on TV watching, where TV crowds out social interactions.)

Other researchers have since studied crowding-out effects with similar results. In 1993 sociologist Bruno Frey found that Swiss citizens, when asked whether they would approve a hypothetical government proposal to site nuclear waste storage in their region, were split about equally on the question. When Frey rephrased the question to include the possibility that the government would pay the citizens for housing the waste, however, they shifted to three to one *against* the proposal. The prospect of hosting a waste dump was twice as unpopular when it was presented as an activity for which the community could be compensated as when it was presented as an issue of civic duty. Later work by Frey and his colleague Lorenz Goette found that in real-world situations where money was offered as a reward for volunteering, it depressed the number of hours of labor the average volunteer contributed. Michael Tomasello, director of the Max Planck Institute for Evolutionary Anthropology, has recently produced experimental evidence that this sort of crowding out can appear in children as young as fourteen months, when an extrinsic reward is tied to an activity they like and then the reward is taken away.

The idea that people behave differently if they are doing something for love or for money won't seem terribly surprising to anyone who's ever had both a job and a hobby, but many in the world of academic psychology regarded Deci's findings as perverse. In 1970, theories of human motivation, as well as the practical use of rewards in the schoolroom and the workplace, were often based on simple ideas of stimulus—adding any new reward to an existing activity would make people do more of it. This framework made little distinction among different kinds of motivation, and the most general-purpose motivator available has always been cash. Deci's conclusion that payment can crowd out other kinds of motivation flew in the face of both existing theory and practice. His experiment and the subsequent research on the crowding-out effect kicked off an academic disagreement that continues today.

In 1994, Judy Cameron and David Pierce of the University of Alberta analyzed the results of dozens of studies that had paid experimental subjects to perform various tasks. Their meta-analysis (as such studies of multiple experiments are called) denied the existence of any such crowding-out effect. Deci and research partner Richard Ryan responded in 1999, pointing out that Cameron and Pierce had included a large number of studies noting that people were more motivated to do uninteresting tasks if you paid them, a result no one disputed. What Deci had examined, rather, was intrinsic motivation for tasks a subject was interested in. Deci and Ryan's own meta-analysis, which excluded boring tasks, again found a crowding-out effect. Cameron and Pierce's second meta-analysis, in 2001, conceded that the crowding out of free choice can occur with the introduction of extrinsic motivations. Nevertheless, Cameron and Pierce remained skeptical that the crowding-out effect mattered much in the real world; their focus was on rewards offered in institutional settings, like schools and workplaces. To them, the crowding-out effect seemed concentrated in areas where people had a high degree of freedom to choose their activity. Cameron and Pierce thus concluded that though the crowding-out effect was real, it was minor. After all, how many places are there where someone's free choice of activities matters much to anyone but the individual?

In an age when our free time and talents are joint resources, the answer is "Everywhere."

AUTONOMY AND COMPETENCE

Deci's framing of intrinsic and extrinsic motivations, and of the crowding out of love by money, illuminates a lot about the creation of Grobanites for Charity. Philanthropies vary greatly in determining where their money goes: how much goes to actual recipients and how much for day-to-day operating expenses, including salaries for the people running the organization. The American Institute of Philanthropies gives a passing grade to philanthropies that use 40 percent of donated money for expenses while giving away 60 percent—not terrible, but not great. Philanthropies that limit their expenses to 15 percent and give away 85 percent are rated excellent. And what of Grobanites for Charity—how much of the money donated by their fellow fans do they take for expenses? Nothing. Zero percent. They draw no salary, and as much labor as possible is donated from Grobanites willing to invest time instead of (or as well as) money. Grobanites for Charity doesn't just happen to be a labor of love; it is designed and incorporated as a labor of love. (The word *incorporated* actually means "embodied"—incorporation is the embodiment of a group's shared efforts and goals.)

Intrinsic motivation is a catch-all label, grouping together several reasons one might be motivated by the reward that an activity creates in and of itself. Deci identifies two intrinsic motivations that might be labeled "personal": the desire to be autonomous (to determine what we do and how we do it) and the desire to be competent (to be good at what we do). In the Soma experiment, the students who continued to play with the pieces during their break were motivated by desires both for autonomy (the work was under their control) and for competence (Soma is a game where continued effort brings improvements in skill). This finding is typical of games. A study of video games concluded that the principal draw for the players was not graphics and gore but the feelings of control and competence the players could attain as they mastered the game.

On the other hand, the group that was paid to assemble the Soma pieces had their intrinsic motivations diminished. Their sense of autonomy was crowded out by the presence of a predictable extrinsic reward. Likewise, the pleasure of competence, once paid for, stopped being a pleasure; when they wanted to get better at solving Soma configurations in order to increase their pay, improving at the same task for its own sake lost enough of its value to depress free choice.

Similarly, Grobanites for Charity and the Josh Groban Foundation don't differ only in terms of contracts and salaries. Every aspect of the two organizations differs, in ways that are tied to preserving the Grobanites' intrinsic motivations. For example, the Groban Foundation doesn't even have its own website; it just has a small section on JoshGroban.com containing brief updates and press releases. The feel is clean, professional, and minimal. The Grobanites for Charity site, on the other hand, doesn't look like that at all. It looks like 1996 threw up, with all of the loopy touches that characterized web design in its early days—lists bulleted by hand-drawn hearts, and colored tabs showing the viewer the sections. It looks, in other words, as if it were done by amateurs, because it *was* done by amateurs, not just in the sense of "not professional," but also in the original sense of the word *amateur*: someone who does something for the love of it.

Markus started designing sites for the Grobanites when she

was a teenager—the original auction site was her first effort—so building the Grobanites for Charity site was a considerable learning experience. Learning on the job may seem opposed to the desire to feel competent, but competence is a moving target. Taking on a job that is too large and complex can be demoralizing, but taking on a job that is so simple that it presents few challenges can be dull and demoralizing. The feeling of competence is often best engaged by working right at the edge of one's abilities. The feeling that I did this myself and it's good, often beats the feeling that Professionals did this for me and it's perfect.

This effect is general. Back in the web's early days, a site called Geocities offered its users personal home pages, on which they could post writings, drawings, pictures, whatever they wanted, for other people to see. At the time it launched, I was running the production department for a web design firm in New York, and I was certain Geocities was going to fail. I'd seen the amount of work that went into designing a usable website, from the navigation to the design to the layout, and I knew that a bunch of amateurs could not even approximate the quality of what professional designers were creating. No one would want to have their own mediocre page, when there was all of this professional work being put on the web at the same time.

I was right about the design quality of the average Geocities page, but I was completely wrong about the popularity of Geocities; it quickly became one of the most popular sites of its day. What I hadn't understood was that design quality wasn't the sole metric for a webpage. Webpages don't just have quality; they have qualities, plural. Clarity of design is obviously good, but other qualities, like the satisfaction of making something on your own or learning while doing, can trump it. People don't actively want bad design—

it's just that most people aren't good designers, but that's not going to stop them from creating things on their own. Creating something personal, even of moderate quality, has a different kind of appeal than consuming something made by others, even something of high quality. I was wrong about Geocities because I bet that amateurs would never want to do anything other than consume. (That was the last time I ever made that mistake.)

MEMBERSHIP AND GENEROSITY

Yochai Benkler, a legal scholar at Harvard, and Helen Nissenbaum, a philosopher at NYU, wrote a paper in 2006 with a mouthful of a title: "Commons-Based Peer Production and Virtue." Commons-based peer production is Benkler's term for systems that rely on voluntary contributions to operate—systems that rely on cognitive surplus. In their piece, they describe the positive characteristics that such participation both relies on and encourages. Like Deci, Benkler and Nissenbaum focus on personal virtues like autonomy and competence. But where Deci's Soma work focused mainly on personal motivations, they spent considerable time on social motivations, motivations that we can feel only when we are part of a group. They divide social motivations into two broad clusters—one around connectedness or membership, and the other around sharing and generosity.

Observing several such participatory examples, including especially software creation through shared contributions among peers (a model called open source software), Benkler and Nissenbaum conclude that social motivations reinforce the personal ones; our new communications networks encourage membership and sharing, both of which are good in and of themselves, and they also provide support for autonomy and competence. Deci's early work on Soma did hold one clue to this effect: verbal rewards for completing Soma configurations like "That's very good" or "That's much better than average for this configuration" produced improvements in performance, improvements that persisted even after the verbal feedback ended. Verbal feedback seems like it should be just another extrinsic reward, like money. When it is genuine, though, and comes from someone the recipient respects, it becomes an intrinsic reward, because it relies on a sense of connectedness.

Social forms of organization can affect even the most seemingly personal issues. Katherine Stone, a U.S. advocate for women suffering from anxiety disorders, noted the recent and rapid growth in postpartum support groups organizing via Meetup.com, a service that uses the internet to coordinate real-world meetings of the like-minded. Stone explained this rapid growth by saying, "Women going through postpartum depression . . . WANT AND NEED TO TALK to other women who are just like them. To share. To see they are not alone. To see they will get well." The motivation to share is the driver; technology is just the enabler.

This feedback loop of personal and social motivations applies to most uses of cognitive surplus, from Wikipedia to PickupPal to Grobanites for Charity. Grobanite donors and supporters get two messages: both *I did it* and *We did it*.

The changed potential for membership and sharing shows up in the design of the Grobanites for Charity website. Now, the design of a website may not seem to have much to do with fostering a sense of membership, but something designed by an amateur can actually create better conditions of membership than a professional design can, in the same way lolcats sends the message *You can play this game too.*

As an analogy, consider the kinds of kitchens you see in photographs in *House Beautiful* and *Better Homes and Gardens,* designed to a fare-thee-well with a place for everything and everything in its place. My kitchen is not like that. (Perhaps yours isn't either.) But if you were a guest at a dinner party, you likely wouldn't dare set foot in a *House Beautiful* kitchen, because the design doesn't exactly scream *Come on in and help!* My kitchen, on the other hand, does scream that—you wouldn't feel much compunction about grabbing a knife and dicing some carrots if you felt like it.

Markus's Grobanites for Charity site works along these lines. It has a less-than-polished design, especially in comparison to JoshGroban.com, but it looks more inviting that way, both figuratively and literally, and this inviting quality is embodied in every aspect of the site. The various links on the main page are pretty much what you'd expect—Donate, Auction, About Us, and so on. And then there's the section called Thank You, which looks like this:

> **A Special "Thank You" To . . .**
>
> . . . Shari, for generously donating her time to make the original sweatshirts, t-shirts, and beanies to raise money for the charities.
>
> . . . Ellen, for donating her portraits of Josh (and David Foster) to raise thousands of dollars for the charities.
>
> . . . Linda, for making our Grobanites for Charity cards and for sending out the Thank You cards to the DFF donors.

It continues in that vein for some time, thanking more than 350 people by name, and the page instructs readers to "Please let us know if we missed anyone!"

Any large effort requires a certain amount of scut work to be done, and the connections exemplified on the thank-you page provide an incentive for people to do jobs not just because they need to be done, but because they become visibly valuable to the group. Rather than functioning as a generic volunteer corps that does whatever is needed, people have an incentive to take on specific jobs. Pollyann Patterson does all the mailings of buttons and magnets, at such scale (Grobanites love buttons and magnets) that she's become known in the community for doing this job, and doing it well. On the About Us page, she's said to handle "Button and Magnet Coordination," and her thank-you reads, "Polly, for mailing out all the buttons and magnets for Grobanites for Charity whenever we need it!" Such recognition is part of the communal glue that lets the Grobanites get on with the harder and larger job of raising money. (As Cicero said two millennia ago, "Gratitude is not only the greatest of virtues, but the parent of all the others.")

Now, this parade of thanks may seem too cute for words, but it's hard to argue with success. Since 2002 the Grobanites have raised more than a million dollars, and they've sent 100 percent of that money on to other charities. Not bad for a bunch of amateurs.

Although the Grobanites' circumstances are unusual, the change in charity isn't. The Grobanites, ahead of the curve, did everything by hand, like taking a generic auction site and turning it into a custom platform for fund-raising; but now several services have sprung up to make setting up sites for charity easier. Facebook hosts an application called Causes that allows users to donate to charitable causes; it lists more than 350,000 causes, which have cumulatively recruited millions of users and received millions of dollars. (Aflac, a cancer center at Children's Healthcare of Atlanta, has more than a million Facebook members.) Many sites are set up to help people donate to various charities, like DonorsChoose.org, for educational causes, or Firstgiving.com, an online platform for nonprofits.

Many other sites are dedicated to helping users donate time and expertise as well as money. NetSquared.org supports nongovernmental organizations working on global aid or development, Idealist.org helps people find opportunities for local community development, and Care2.com supports environmental initiatives. The microlender Kiva.org uses donations from individuals as capital for providing loans to people in the developing world. In 2008 its fund-raising was so successful that it outstripped the site's ability to vet potential recipients; Kiva ran out of projects to offer loans to and had to turn donors away. And some people donate their time to charity directly, like users of Extraordinaries at BeExtra.org, who use their mobile phones to help label photos for everything from museum exhibits to documentation of environmental degradation. These novel modes of charity rely not only on the existence of tools that connect us and let us volunteer our time, talents, or money; they rely on our being motivated to do so as well.

AMATEUR MOTIVATION, PUBLIC SCALE

Amateurs are sometimes separated from professionals by skill, but always by motivation; the term itself derives from the Latin *amare*—"to love." The essence of amateurism is intrinsic motivation: to be an amateur is to do something for the love of it. This motivation also affects how amateurs work in groups. Keeping a large group focused can be a full-time job. (It's middle management's reason for being, in one phrase.) Organizing groups into an effective whole is so brutally difficult that, past a certain scale, it requires professional management. Professional managers in turn require salaries, and salaries require income and bookkeeping and all the rest of the trappings of a formal organization, meaning there is a huge step between a bunch of people who really care about some issue and an organization of people who really care about that issue and work together to do something about it.

As always, high hurdles to an activity reduce the number of people who do it, and the hurdle of large-scale coordination has largely separated amateurs from professionals. People doing something for the love of it, whether it's collecting donations or making music or engaging in a hobby, usually do it in relative obscurity; church basements, public libraries, rec rooms, and garages tend to be the homes of amateur groups. Professional activities can be more publicly visible (and indeed, many professional groups seek public visibility, whether in the market or the media, as an explicit goal). This has accustomed us to two modes of behavior: people who act from intrinsic motivations—amateurs—operate in relatively private circumstances, while people who act from extrinsic motivations operate in more public ones.

What we're seeing now, though, is that amateur motivation and private modes of behavior weren't really linked at all. Instead, the old constraints on organized action forced these phenomena together mostly as a negative option: when pursuing an intrinsic goal in public required considerable work, the amateurs largely opted out of public action. We have always wanted to be autonomous, competent, and connected; it's just that now social media has become an environment for enacting those desires, rather than suppressing them. An all-volunteer charity group meeting in a church basement used to have different access to the public sphere than a professional philanthropic organization, but that all-volunteer group's website is now as accessible as the professional organization's—GrobanitesForCharity.org is as easy to get to as DavidFosterFoundation.org. The site created by an amateur may not attract as many visitors as the one created by a professional but a key obstacle separating the amateurs from the professionals has been removed.

Back when coordinating group action was hard, most amateur groups stayed small and informal. Now that we have tools that let groups of people find one another and share their thoughts and actions, we are seeing a strange new hybrid: large, public, amateur groups. Individuals can make their interests public, more easily, and groups can balance amateur motivation and large coordinated action more easily as well.

The geographic range of collaborative efforts has spread dramatically. When Linus Torvalds first asked for help creating what would become the Linux operating system, he received only a few replies, but they came from potential participants all over the globe. Similarly, Julie Clarke, Valerie Sooky, and Meg Markus all lived in different places when they were forming Grobanites for Charity, but that didn't stop them from creating a charity that's raised a million dollars.

We're used to the word *global* meaning "really big"—global corporations are larger than national ones, global markets have more participants than local ones, and so on. But this, too, was just a side effect of difficulty—when big organizations were hard to run, and increasingly harder to run the bigger they got, an organization had to get very big before it could "go global" (in MBA-inflected parlance). But globalization isn't necessarily about size; it's about scope. Now that the difficulty of coordinating interactions has fallen, it is entirely possible to have a tiny global organization. Geography still matters, but it is no longer the main determinant of participation.

Japanese anime (animated cartoons) are often subtitled in English by networks of volunteer fans, in a process called fansubbing; fansubbing networks are small and global by nature, and different groups of fansubbers typically concentrate on particular anime shows or artists. Yahoo.com hosts a mailing list for sufferers from Crohn's disease, providing a place for them to share their worries and observations; it has a few hundred active members, drawn from Europe, North America, and Asia. The online craft bazaar Etsy also has merchants from all over the world. In 2008 the United States changed its laws to require more chemical testing of clothing and toys (the Consumer Product Safety Improvement Act of 2008, or CPSIA); ordinarily, such a change would force each craftsperson to look around for information on CPSIA, but Etsy provides not just a virtual storefront for its vendors but a virtual water cooler as well. Vendor forums erupted with conversations like "EVERYTHING you need to know about the CPSIA" and "CPSIA testing—making it affordable for you." These conversations attracted participants in England, Wales, Canada, and Australia as well as the United States. None of the participants had lawyers skilled in trade compliance issues—almost all were individual craftspeople without lawyers at all, but even spread across the globe, they were able to work together to quickly understand the law's intent and effect, something they couldn't have done without the social media.

FEEDBACK LOOPS

If intrinsic motivations are fundamental to human nature, and if satisfying them satisfies us, then the use of tools that satisfy those motivations should spread. In particular, if the social media provides a platform for creating and sharing at a low enough cost, then participation in activities that reward an intrinsic motivation should rise, even if the satisfaction lasts only a brief moment. This is just what's happened. One of my students, Victoria Westhead, documented the rise of something she labeled digital folk art—the amateur production of words, sounds, and images designed to be engaging or amusing and intended for amateur circulation rather than inclusion in any formal publication. Digital folk art often takes the form of a mashup, the combination of existing materials into something new. (Lolcats is an example of a mashup: a person adds a caption to an existing picture.)

Digital folk art has existed almost as long as computers have. The original form was ASCII art, dating from the mid-1960s, before computers had screens. ASCII (the American Standard Code for Information Interchange) describes the way computers print letters and numbers; ASCII art was made by printing them in such a way as to form black-and-white images that would materialize when viewed from a distance. Digital folk art spread as the web did; many animations simply ran alongside music, with no real synchronization—a dancing baby, dancing hamsters, contests to see who could best alter famous movie scenes. These artworks were circulated via e-mail and other social tools with no goal other than to share something funny, yet many of them eventually reached audiences in the millions.

The spread of digital hobbies hardly seems significant, in part because we've learned to regard amateur interests as faintly ridiculous, if not actively suspect. While I was growing up in the United States in the 1970s, I learned, without being explicitly told, that grown men who built model trains or women who created macramé were, in some unstated way, pathetic. Meanwhile it was perfectly acceptable to spend hours every day watching *The Partridge Family* and *The Brady Bunch* (a task I performed, as most of us did, as if it were my job).

Despite my teenage attitude toward other people's interests—hobbies are for losers, but TV is for everybody—hobbyists and crafters continued to work in basements everywhere. Their motivations were largely personal: a desire for autonomy and competence. Social motivations—desire for membership and sharing—were less prominent, simply because what economists call "discovery costs" were high; it was hard for hobbyists to figure out who else nearby liked model trains or macramé. Indeed, as Harvard sociologist Robert Putnam documented exhaustively in his 1995 book *Bowling Alone: America's Declining Social Capital,* postwar America saw a general decline in social connections in an incredibly broad array of areas, from number of close friends to participation in hobbyist groups to participation in amateur sports leagues (as suggested by Putnam's title). Putnam argued that this reduction in social capital was driven by suburbanization, longer commute times, and increased television watching.

If the only thing our new communications tools allowed was

the release of pent-up desires, the effect would have been like a cork popping; satisfaction of our latent needs for autonomy and competence would pour out quickly and then stabilize at some new level. But that's not what's happening. The flow of amateur production and organization, far from stabilizing, continues to increase, because the social media rewards our intrinsic desires for membership and sharing as well.

Broadcast media, like television, clearly filled some human needs, but those needs that they couldn't fill well became harder to see and, ultimately, harder to imagine. Now those desires are starting to reappear because the social media has made them both expressible and visible, and also because personal motivations and social ones amplify each other in a feedback loop. Indulgence in feelings of membership and sharing can increase our desire for more connectedness, which increases its expression, and so on. Someday we will reach a new equilibrium of amateur participation versus consumption, but since large-scale joint projects and shared spaces didn't really exist in the twentieth century, we've got a long way to go before we get to that equilibrium.

Social media also drives discovery costs through the floor: web access allows you to find other people who like building model trains and doing macramé, or designing paper airplanes, or dressing up as anime characters, or practicing jnana yoga, or knitting socks, or photographing pay phones, or cooking Catalan food, on and on, at any hour of the day or night, worldwide. As my NYU colleague Nicholas Mirzoeff has remarked, the reason the web continues to astonish is simple: "The web means we're finally being exposed to the full crazy range of what people are actually interested in."

People who care passionately about something that seems unimportant to the rest of us are easy to mock. The satirical publication *The Onion* sometimes runs opinion pieces by a nerdy know-it-all named Larry Groznic, who defends sacred works of geek culture. The headlines alone read like a compendium of obsessions: "When You Are Ready To Have A Serious Conversation About Green Lantern, You Have My E-Mail Address"; "I Appreciate The Muppets On A Much Deeper Level Than You"; "Now More Than Ever, Humanity Needs My *Back to the Future* Fan Fiction." Part of the joke is that the internal concerns of any particular community appear picayune to the outside eye; but to be a member of a community of shared interests is to care, deeply and in detail, about things the general public doesn't spend much time thinking about. If you want to see this effect in action sans Larry Groznic, go to a newsstand and buy a magazine on a subject you care nothing about. If you read *Vogue,* get *Guns and Ammo;* if you read *Golf Digest,* pick up *Tiger Beat;* and as you read, imagine what someone who liked that magazine would think about your interests.

In a world of high discovery costs, people who really cared about the Muppets or wrote *Back to the Future* fan fiction had a hard time finding others who shared their interests, and without anyone to converse with, learn from, or show off for, they weren't likely to fully express those interests. In a world of low discovery costs, however, people who care about certain things can find each other and interact, away from the mass of us who just don't get it.

Amateurs generally use public access not to reach the broadest possible audience but to reach people like themselves, just as the Grobanites for Charity tried to reach more Groban fans. It's possible to give money to Grobanites for Charity without being a Josh fan, but it's certainly not the normal case. It turns out that having access to everyone is now an easier way to reach Groban fans than using more targeted means. (*Targeted* here refers to the ways in which ordinary people are, in the worldview of broadcast media, targets.) This "go public to find people who think like you" strategy has created an unprecedented increase in the amount of material that is available to the public but not intended for the public—its creators are looking not to reach some generic audience but rather to communicate with their soul mates, often within a sense of shared cultural norms that differ from those of the outside world.

Consider the denizens of FanFiction.net, the community of people who write new stories set in the imagined worlds of existing fictional works. The most fecund of these communities are people writing stories set in the Harry Potter universe—FanFiction.net hosts more than half a million Potter stories (and still more appear on sites like FictionAlley.org and HarryPotterFanFiction.com). Hundreds of the stories run to over one hundred thousand words, roughly the length of J. K. Rowling's original novels. FanFiction.net doesn't just aggregate stories; it hosts a community in constant conversation with itself. If "thank you" is the coin of the realm among Grobanites, attention is the coin for fan fiction; the plea to "please read and review my story" is so common it has been shortened to "R&R."

Like all communities, the world of fan fiction sometimes gets roiled by violations of its cultural norms. In the Harry Potter community, a fanfic author with the pen name of Cassandra Claire was accused of copying passages into her fan fiction from two books by the fantasy author Patricia Dean. It may seem odd that a group of people publicly engaging in wholesale copyright violation are concerned with plagiarism, but they are, and deeply so. Failure to give credit where credit is due is *the* crime in this community, a violation not of property rights but of deeply held ethical norms about credit. Some fan fiction writers even use a "legal" disclaimer at the beginning of their works, with "legal" in quotes because the disclaimers read like this, misspellings and all:

> "Disclaimer: I don't own these characters, but I do own their personalities! [grin] . . . kind of? I dunno. But anyway, JK Rowling is amazing."
>
> "Disclaimer: i do not own harry potter this is purely a fan written story."
>
> "Disclaimer: Harry Potter Universe in not mine, just Dana Cresswell is :)"
>
> "Disclaimer: I do not own Harry Potter or any of the other characters . . . I just am borrowing them!"

Lawyers would laugh till coffee came out their noses at the idea that writers can legally borrow other writers' characters, that fan fiction is a special class of creativity, or that writers can own new characters or plots in existing fictional universes without the permission of those who created those universes. Even the writers of the disclaimers are unsure about them, like the author who claims to "kind of" own the personalities of characters Rowling invented. Like children staging a wedding, the disclaimers mimic an existing form of obligation while remaining legally inert. They aren't worthless, but their worth lies elsewhere.

The internal logic of the fanfic community becomes clearer in light of the other charge leveled at Cassandra Claire; she was accused of profiteering, which, in the culture of fan fiction, means trying to make money from her fanfic. This was held up as still more evidence that she was impure of heart. Fanfic disclaimers express the logic of giving public credit ("JK Rowling is amazing"), albeit in the language of ownership. This is a "two worlds" view of creative acts. The world of money, where Rowling lives, is the one where creators are paid for their work. Fan fiction authors by definition do not inhabit this world, and more important, they rarely aspire to inhabit it. Instead, they often choose to work in the world of affection, where the goal is to be recognized by others for doing something creative within a particular fictional universe. A robust communal infrastructure is essential to that mutual recognition. Indeed, one of the most lamented effects of *l'affaire* Claire was that it created a schism in the Harry Potter fanfic community.

Seen in this light, it doesn't matter whether the fan fiction authors understand that what they are doing is illegal. By publicly disavowing ownership of JK Rowling's work—something that was never in dispute—they are demonstrating their respect for the source of material that is now integrated into their imagination. They are also carving out a practical distinction between the world of money and the world of love, because even though that distinction is meaningless in a court of law, it is meaningful to them. Within that community, purity of motivation inside the community matters more than legality of action outside it.

INTRINSIC MOTIVATION, PUBLIC ACTION

If you had a few spare weeks to kill, you could spend as much time as you like reading various public utterances on mailing lists, blogs, social networks, wikis, bulletin boards, and every other place online where an individual can, with three minutes of typing and one press of a button, make his thoughts globally available. And if you tried it, you'd get exhausted without coming anywhere close to exhausting what's out there. Indeed, you'd be outstripped by the desire of the world's participants to avail themselves of these newly public channels. No matter how much time you devoted to reading, watching, and listening, the world's amateurs would, in that same period, produce more material—vastly more—than you'd have taken in. By the end of 2009 an average of twenty-four hours of video were being uploaded onto YouTube every minute; Twitter receives close to three hundred million words a day.

When you see people acting in ways you don't understand, you may ask rhetorically, Why are they behaving that way? A better question is this: Is their behavior rewarding a desire for autonomy, or for competence? Is it rewarding their desire to feel connected or generous? If the answer to any of those questions is yes, you may have your explanation. If the answer to more than one of those questions is yes, you probably do.

People's intrinsic motivations are strong enough that they gravitate toward experiences that reward them. They also sometimes act out when they are offered phony versions of the participation. Back in 1998, *People* magazine's website asked its readers to rank a list of the 50 Most Beautiful People of that year, a list chosen but not yet sorted by the site's editors. In an online poll they requested that their readers help them separate the most beautiful from the merely fabulously beautiful.

It's hard to imagine a more cynical way to involve readers. The editors were confident that the participants would internalize their publication's view of the world and produce a list—with Leonardo DiCaprio first, then Kate Winslet, or whatever—as acceptable to their audience as a list they could create themselves. The polling effort was a fairly transparent attempt to drive traffic (when we are not targets or eyeballs, we are traffic). But *People*'s otherwise hermetically sealed chamber of beauty had one tiny crack, and that was the write-in vote. With this option, *People*'s traffic (which is to say, us) could participate without choosing from the proffered alternatives. Write-in votes are usually a way to let individuals blow off steam without changing an election's outcome, since voters cannot easily organize enough support around a particular write-in candidate to challenge the listed ones.

Or rather, once upon a time they had no such easy way. But by 1998 the web offered new tools for group synchronization. So began the campaign to anoint Hank, the Angry Drunken Dwarf, as People.com's Most Beautiful Person of 1998.

Hank, the Angry Drunken Dwarf, née Henry Joseph Nasiff, Jr., was pretty much what his moniker described. He'd been an occasional guest on the radio show of shock jock Howard Stern; famous for his inebriated rage, he'd played the role for about two years when the *People* poll launched. Nasiff was a kind of funhouse version of the people of *People*, notable for his appearance and demeanor but in a manner utterly unlike DiCaprio or Winslet. The write-in campaign for Hank was started by Kevin Renzulli, creator of a fan site, KOAM.com, devoted to Stern, the self-described King of All Media. Stern then took the idea from Renzulli and rebroadcast it; the denizens of various discussion boards and mailing lists then picked it up, and all these novel forms of media generated a sudden surge of votes for Hank.

Having to act on behalf of an authority can be one of life's great demotivators, and given the chance, people will often do something other than what is expected of them out of sheer defiance. (No one with children has missed this phenomenon.) Voting for Hank offered people a chance to violate *People*'s expectations while still playing by its rules. Hank won by a landslide, with nearly a quarter of a million votes; second place was the professional wrestler Ric Flair, another write-in. DiCaprio, the first of *People*'s original choices, came in third, with a mere fourteen thousand votes.

The lesson of Hank, the Angry Drunken Dwarf is this: If you give people a way to act on their desire for autonomy and competence or generosity and sharing, they might take you up on it—every successful example in this book involves harnessing those intrinsic motivations in one way or another. However, if you only pretend to offer an outlet for those motivations, while actually slotting people into a scripted experience, they may well revolt.

We used to pursue our intrinsic motivations in private, either alone or among family and friends. The link between intrinsic motivation and private action was never deep, however. Back when entering the public arena was hard—like taking a separate job—most of us simply didn't bother. Loose collections of amateurs may have been willing to try to accomplish things in public, but the organizational hurdles were too high. Now the barriers are low enough that any of us can publicly seek and join with like-minded souls. The means for harnessing our cognitive surplus are the new tools we have been given, tools that both enable and reward participation. Our motivations for using those tools are the ancient, intrinsic ones, motivations previously remanded to the private sphere but now bursting out in public. To be turned into something real, however, all of this raw capability still requires opportunity.

4

Opportunity

All of us have intrinsic motivations, desires to do things because we enjoy them in and of themselves. We now have tools for communicating and sharing, new means for indulging ourselves in those motivations. Means and motive aren't enough to explain what's happening with the new uses of our cognitive surplus, however. We also have to account for opportunity, ways of actually taking advantage of our ability to participate in concert where we previously consumed alone. The cognitive surplus is not simply trillions of hours of free time spread across two billion connected individuals. Rather, it is communal; we must combine our surplus free time if it is to be useful, and we can do that only when we're given the right opportunities.

We create opportunities for one another that we wouldn't otherwise have. By treating one another well (fairly, if not always nicely), we can create environments where the group can do more than the individuals could on their own. (The Grobanites couldn't have raised a million dollars if they lacked the means to pool their resources or the motivation to offer communal appreciation as a reward for hard work.) Human character is the essential component of our sociable and generous behaviors, even when coordinated with high-tech tools. Interpretations of those behaviors that focus on the technology miss the point: technology enables those behaviors, but it doesn't cause them.

To take one example, in the 2000s innumerable stories were written about how seniors were actually starting to use social tools, stories with titles such as "Old People Like the Internet," "Geezers Need Excitement: What Happens When Old People Go Online?" and "More Older People Turning to the Internet to Find Love." These stories always had a surprised feel about them. Many people assumed in the mid-1990s that no one over fifty would want to adopt computers and networks. In the 1990s, when anyone asked, Will older people adopt all this confusing new technology? the assumed answer was no, but that was the wrong answer, because it was the wrong question. The right question was, Will older people adopt new ways of communicating with friends and family?

No one wants e-mail for itself, any more than anyone wants electricity for itself; rather, we want the things electricity enables. Similarly, we want the things e-mail enables—news from home, pictures of the kids, discussion, argument, flirtation, gossip, and all the mess of the human condition. The surprise behind those "Old people communicating with each other!" articles came from a focus on the technical means rather than on the social opportunities of that communication.

When some surprising new thing happens, we want an explanation, and we usually turn to the one that hinges on the novelty.

If people are using their surplus time and talents in generous and public ways, then we assume the cause is new tools: the web, mobile phones, new software, everything that didn't exist in the past. These kinds of technology-centric observations locate the surprise in the new tools. There is another possibility, though. When a surprising new thing happens, instead of asking *Why is this new?* we can ask *Why is it a surprise?*

Many of the unexpected uses of communications tools are surprising because our old beliefs about human nature were so lousy. Daniel Kahneman, the first noneconomist to win a Nobel Prize in economics, for his work on ways in which humans aren't rational economic actors, calls this effect "theory-induced blindness": adherence to a belief about how the world works that prevents you from seeing how the world really works. (This is the mechanism of the "milkshake mistake" in Chapter 1.)

As we have seen, the question, Why are all these people working for free? presupposes a theory of human action based mainly on personal and financial motivation: The sensible reason to do things is for money, so doing things for free requires a special explanation. Within that theoretical framing, there is no good reason why someone would upload their video to YouTube or edit a Wikipedia article. The problem here isn't with the behaviors, it's with the explanation. Once you stop asking why people do these things "for free" and just start asking why they are doing them, the whole range of intrinsic (and nonfinancial) motivation becomes part of the explanation.

Our collective surprise about older people using social tools had little to do with technology and much to do with this kind of blindness. A surprise is not just about having new information; we integrate new information about the world all the time. Yesterday it was raining and today it's not; that's new information, but it's not a surprise. A surprise is having new information that violates our previously held assumptions. A surprise, in other words, is the feeling of an old belief breaking. The surprise here was that our assumptions about how off-putting new communications technologies were turned out to be absolutely worthless.

Many of the stories we tell ourselves about the tools we use are really stories about human motivation. We grossly overestimated the degree to which e-mail would always seem futuristic and hard to use, we grossly underestimated the technical talents of older people, and we simply ignored the basic truth of technology: if a tool is useful, people will use it. (Surprise.) They will use it even if the tool is very different from what existed before, provided it lets them do things they want to do. The mystery isn't why older people started e-mailing each other; the mystery is how we could have convinced ourselves that e-mail use was mainly about technological novelty rather than social continuity.

Given the right opportunities, humans will start behaving in new ways. We will also stop behaving in annoying old ways, even if we've always tolerated those annoying behaviors in the past. At the 2006 Burning Man, an annual cultural festival in the Nevada desert, some energetic techies built a phone booth that could make calls via satellite. They proudly invited their fellow Burners (as attendees are called) to make free phone calls, but few people were able to, because no one could remember any phone numbers, and no one had brought their mobile phones to look them up. (Mobile phones don't work in that part of the desert.) For the better part of a century, the telephone system required us to memorize strings of numbers if we wanted to talk to our friends, so we did. But memorizing numbers turned out to be nothing more than a frozen accident. We memorized phone numbers when we had to, but we never liked memorizing those numbers and we were never very good at it. We did it because it was a requirement for other things we did like, such as talking with our friends. The minute phones provided us with speed dials and address lists, that accident unfroze and melted away.

Many of our behaviors are like memorizing phone numbers, held in place not by desire but by inconvenience, and they're quick to disappear when the inconvenience does. Getting news from a piece of paper, having to be physically near a television at a certain time to see a certain show, keeping our vacation pictures to ourselves as if they were some big secret—not one of these behaviors made a lick of sense. We did those things for decades or even centuries, but they were only as stable as the accidents that caused them. And when the accidents went away, so did the behaviors.

Every surprising bit of new behavior described here has two common elements: people had the opportunity to behave in a way that rewarded some intrinsic motivation, and those opportunities were enabled by technology but created by human beings. Those bits of new behavior, though, are extensions of, rather than replacements for, much older patterns of our lives as social creatures.

SKATEBOARDS AND EASELS

In the early 1970s drought and a recession left many swimming pools in southern California dry. An empty swimming pool doesn't seem like it would be of much use to anyone, but in the town of

Santa Monica a group of kids took advantage of the recession-drained pools. This group, the self-styled Z-Boys, began riding on skateboards inside the empty pools, running up and down the curved walls and across the bottoms. The pools enabled a remarkable range of new tricks, as skating up one wall provided enough momentum to speed back down again and across the bottom to the other wall, where they could repeat the process. With just a few kicks and an assist from gravity, a skateboarder could skate in a style that was dramatically faster and more athletic than anything he could do on a street or sidewalk. As they skated (or sat around nursing injuries), the Z-Boys exchanged tips and tricks; they continually tested new ideas and adopted, improved upon, or discarded them. The most daring move involved starting at the top of a pool and racing straight down the side: a skateboarder with perfect posture and a few well-placed accelerating kicks could generate enough momentum to skate up the opposite wall and back out of the pool, coming to rest at the top.

The spread of these techniques was driven by spirited competition. We often think of competition as pure conflict, the way firms compete in a market, happy to drive one another out of business. In groups of people who know one another and share the same interests, though, competition can take on a collaborative quality. The Z-Boys competed not to end the development of skating technique but to extend it. Instead of trying to come to some final or right way of skating, or to master some hidden and uncopyable technique, they developed new styles and tricks out in the open, challenging in order to invite a response.

Eric von Hippel is an MIT economist who studies "lead user innovation," innovation driven not by a product's designers but by its most active users. He often uses the Z-Boys to illustrate a more general fact: that a tool's capabilities don't completely determine its ultimate functions. Instead, users can press a tool into service in ways that the designers never imagined, and those new functions are often discovered and perfected not by a burst of solo inspiration but by exploration and improvement among a collaborative group.

Much of modern skateboard culture came out of the Z-Boys' collaboration. In nearby Del Mar in 1975 they appeared in their first formal competition—and their mastery of their style was so complete that they made up half the finalists. Within months their athletic mode and airborne moves were on their way to becoming the new norm, first nationwide and then worldwide.

That network of California skateboarders created an environment where people who liked skateboarding could get better at it together. A sense of membership, of belonging to a group that is animated by a shared vision or project, can spark a feedback loop in which autonomy and competence improve as well. People who are part of a network where they become better at something they love tend to stay. As the group's ability to learn and work together gets stronger, it attracts more participants. The newcomers who don't become part of the core group often take the ideas out to the wider world. The Z-Boys' style of skating didn't become global because everyone became a Z-Boy; rather, peripheral members of the group became ambassadors for those ideas.

In *Collaborative Circles: Friendship Dynamics and Creative Work,* Michael Farrell details how groups of friends and collaborators both improve the ideas of a group and spread them. He details several case studies, starting with the group of French painters known as the Impressionists. The core members of this group, Claude Monet and Auguste Renoir, met when they were studying at the painting studio of Charles Gleyre. The group later met weekly at Café Guerbois, slowly expanding to include Edouard Manet, Edgar Degas, Berthe Morisot, and Camille Pissarro. (Morisot, a woman, couldn't come to the gatherings because cafés were off-limits to women; her inability to participate made her a second-class citizen, illustrating how opportunities are affected by social structures, whether positively or negatively.)

Like the Z-Boys, the core group of Impressionists provided an environment in which new ideas could be quickly tried and improved or discarded; as other participants came and went, taking the ideas with them, that milieu served as a point of diffusion for the group's ideas. As Farrell puts it, "Most collaborative circles consist of a core group who interact frequently and a peripheral 'extended' group who vary in their degree of involvement. The core comprises those members who meet together on a regular basis, discuss their work, and through their interaction develop a new vision," while the extended group disseminates the ideas arising from the core.

The high culture of French Impressionism and the skateboarding culture of the Z-Boys had similar dynamics. The hothouse environment of a collaborative circle can make the ideas and achievements of the participants develop faster than if the participants were all pursuing the identical goals without sharing. Our ability to simultaneously pursue our own goals while being mindful and supportive of other people's goals is fundamental to human life—so fundamental, in fact, that we actually have trouble turning it off.

ULTIMATUM GAME

For the second half of the twentieth century, mainstream economics (often labeled neoclassical economics) typically located the effects of human emotion outside financial transactions, either before or after. These economists thought of emotion as the incentive for a transaction ("He's buying new shoes because he'd like to be better dressed") or as the result of a transaction ("He's happier now that he has new shoes"). The transaction itself was generally assumed to be bloodless. That simplifying notion is at odds with much of real life. It is also at odds with a large and increasingly important body of social science.

The vast and still-growing body of experimental cognitive science called behavioral economics is demonstrating that humans don't always act in self-interested ways, and that transactions themselves have an emotional component. Behavioral economics often tests ideas by involving subjects in games. By creating a simple set of rules, experimenters can observe human behavior under relatively controlled conditions. The trick is to create an experiment that will cleanly illuminate the part of human behavior the researcher is interested in. One of the most famous and elegant experimental games in social science is called the Ultimatum Game, first tried in 1982 by Werner Güth, Rolf Schmittberger, and Bernd Schwarze at the University of Cologne, and repeated countless times around the world, because its results were so at odds with the predictions of neoclassical economics.

The Ultimatum Game is a two-person interaction. Imagine you and a stranger are the players, and you each have a role: your unknown partner is the proposer, and you are the responder. The game starts when the researcher gives your partner ten dollars, instructing her that she is to decide how she would like to split the money between the two of you. Once she proposes such a split, you cannot change it. The only say you have in the matter is to decide whether to accept or reject her offer. (Hence the game's name.) If you accept, she gets to keep her proposed piece of the ten dollars, and you get the remainder. If you refuse, though, neither you nor she gets anything.

Neoclassical economics predicts the outcome of this game quite clearly: she will propose keeping nine dollars for herself, offering you a dollar, and you will accept, because you will be a dollar better off than if you refuse. One dollar, however small it is relative to her share, is still more than nothing, and hence preferable to nothing.

So the theory goes, but the game doesn't run this way in practice. Instead, the proposer typically offers amounts between four and five dollars, which the responder generally accepts. When the proposer does offer lower amounts, the responder typically refuses, costing both the participants any reward. The lower the amount offered, the greater the likelihood of refusal. This result—fairly intuitive, if you imagine yourself on the short end of that particular stick—was a shock to neoclassical theory (what rational actor would give up a free dollar for the sake of mere emotional satisfaction?). As the results of the Ultimatum Game became more widespread, so did the challenges to its conclusions.

Versions were run with hundreds of dollars at stake, with ever-tighter controls on the anonymity between the participants so they wouldn't worry about retribution, with experimental subjects of different ages, different classes, and different cultures. In one version called the Dictator Game, the proposer is able to declare the terms of the split without the recipient's having any say at all. Even here, the proposed split was more generous than expected. The experiment was performed in countless variations, but the attempt to uncover the secretly rational core of humanity simply failed. No group of proposers playing the classic version of the Ultimatum Game ever behaved as selfishly as neoclassical theory predicted, and no group of responders was ever moved to accept a proposed split that deviated too far from some perceived sense of fairness, no matter how sensible such a choice would be in the short term.

In the Ultimatum Game, people behave as if their relationship matters, even if they are told it doesn't, even if they are assured it doesn't, even if they have only a single interaction with an unknown partner. Critiques of the Ultimatum Game have contended that the game would work if the participants were convinced that their action would have no social consequences outside the game. This criticism misses the fact that if we have a hard time imagining situations in which our dealings will be completely anonymous, it may be because we are deeply social. We are terrible at acting as if we were purely isolated because such isolation is rare and unnatural. (Even economics students, famously among the greediest proposers in the Ultimatum Game, never get anywhere near nine-to-one splits.) Conceived as a psychological rather than an economics experiment, the Ultimatum Game and its variants show that we are incapable of behaving as if we weren't members of a larger society, and as if we didn't gauge the effects of our actions with membership in that society in mind.

Rejecting ungenerous splits, it seems, is a communicative and social act, rather than a simple cognitive mistake. In a variant on the game that strengthens that hypothesis, the proposer is a computer, and critically, the respondent knows the proposer is a computer. Here the responder typically takes the money on offer, since there is no human to punish and no feeling of satisfaction in retaliating against a machine, which wouldn't be able to understand the anger implicit in refusal. Another version was played while the participants were undergoing brain scans; responders who refused ungenerous proposals had increased activity in the dorsal striatum, which is involved in experiencing satisfaction, suggesting that we find keeping defectors in line rewarding, and we will be willing to give up other rewards (in this case monetary) in order to have that feeling.

One of the few variations of the Ultimatum Game that does produce behavior in line with neoclassical predictions is when several responders compete to get a share of a single proposer's cash and do so without communicating with one another. In this case, the proposer can get away with a nine-to-one split, because responders who don't accept such a split get nothing. This is the game played as a market, in which the proposer captures the advantages of pure competition. (The splits shift to one-to-nine when many proposers compete for a transaction with a single responder.) The lesson here is that markets can work as advertised, but they have to be designed and implemented to defeat social coordination.

This is a different, and considerably more limited, finding than the idea that markets are a (or even the) normal case in human life. Indeed, one of the great bulwarks of ethical standards in a society is people's willingness to punish others who defect from norms of fairness or good behavior, even when meting out that punishment costs them something. This is exactly what responders do in the Ultimatum Game when they reject low offers, and because society enjoys the benefits of this individually costly behavior, it is called altruistic punishment. People derive pleasure from punishing wrongdoing, even if it costs them time, energy, or money to do so. In the Ultimatum Game, responders punish skinflint proposers by refusing the offered share of the money, but what they get in return is the satisfaction of knowing the proposer hasn't gotten away with an unfair share.

Market pricing may seem inherently incompatible with communal sharing as a way of organizing human affairs, and it's easy to assume that the less market-oriented a given culture is, the greater the likelihood that its members will be reflexively generous and open with one another. As a way of testing this hypothesis, the Ultimatum Game has been tried in a variety of different cultures, and it turns out that selfishness and market forces are indeed correlated. The surprise is that they are correlated in the opposite way you might expect. Markets support generous interactions with strangers rather than undermining them. What this means is that the less integrated market transactions are in a given society, the less generous its members will be to one another in anonymous interactions.

Far from being incompatible with communal sharing, exposure to market logic actually increases our willingness to transact generously with strangers, in part because that's how markets work. When I am selling something, the economic nature of the transaction actually erodes my interest in how (or whether) I know the buyer. The market acquaints people with the utility of making transactions with people you don't know and with the idea, however implicit, that those transactions are an appropriate way of interacting with strangers.

Just because the norms involved in social production have antecedents in market culture doesn't mean that the two modes can be easily hybridized, though. In fact, switching from paying professionals to create something to having communities do it for the love of the thing may be technically trivial but socially wrenching. There is a constant debate around the donation of blood, plasma, and organs as to whether they should be treated as a communal good or a market commodity. Both methods have been tried in various places, and each has its advantages and disadvantages. But the heat of the debate isn't about marginal differences between Red Cross blood drives (which rely on communal logic) and people selling their blood for plasma (organized in a market). The conflict instead is about the morality of the market as a way to get people to offer their blood or organs. (You can find similar arguments about everything from surrogate motherhood to ticket scalping.)

COMBINABILITY

Information can now be made globally available, in an unlimited number of perfect copies, at zero marginal cost. As a result, every mode of communication that once had to rely on market pricing can now have an alternative that relies on open sharing. (Access to *Encyclopaedia Britannica* uses market pricing, while access to Wikipedia is open; software has undergone a similar shift between commercial and open source versions.) Similarly, the old limitations of TV, radio, and print created a class of media professionals with privileged access to public speech, but public speech can now rely on wide participation. (You have to be hired to be on the nightly news, but not to blog every night.) Many coordination hurdles that required professional managers to direct paid workers can now have an alternative that relies on massively distributed cooperation among amateurs. (Microsoft has to hire and manage the people who create Windows; the group of programmers that created Linux doesn't.)

The change is a matter of simple fact—digital networks make sharing cheap and potential participation nearly universal. But the reaction to that simple fact has often been disbelief or horror, at least from the beneficiaries of market pricing and authority ranking. Maxine Hong Kingston's observation about the wondrous button that says "Publish" upended five hundred years of assumptions that amateurs couldn't share things directly with one another, at least not at large scale.

Alexis de Tocqueville, the eighteenth-century historian, would have understood the advantages of the cognitive surplus. In his book *Public Associations in Civil Life,* he wrote: "In democratic countries, knowledge of how to combine is the mother of all other forms of knowledge; on its progress depends that of all the others." Social production increasingly relies on de Tocqueville's "knowledge of how to combine."

Societies with markets provide people with the experience of interacting with strangers, experience that's needed to take advantage of the cognitive surplus. The trick is in knowing when markets are an optimal way of organizing interactions and when they are not. In particular, as the Ultimatum Game shows, when concerns of fairness around issues other than price arise, normal people's internalized sense of how to treat others is hard to suppress and easy to trigger. As the economist and Nobel laureate Elinor Ostrom has shown, when we assume people are principally selfish, we design systems that reward selfish people. In her 1990 book *Governing the Commons: The Evolution of Institutions for Common Action,* Ostrom characterizes the assumptions that go into these systems:

> When individuals who have high discount rates [i.e., who value present payoffs much more highly than future ones] and little mutual trust act independently, without the capacity to communicate, to enter into binding agreements, and to arrange for monitoring and enforcing mechanisms, they are not likely to choose jointly beneficial strategies.

Assumptions that people are selfish can become self-fulfilling prophecies, creating systems that provide lots of individual freedom to act but not a lot of public value or management of collective resources for the greater public good. Systems designed around assumptions of selfishness can also crowd out solutions that could arise when people communicate with one another and enter into agreements that they jointly monitor and enforce. Conversely, systems that assume people will act in ways that create public goods, and that give them opportunities and rewards for doing so, often let them work together better than neoclassical economics would predict.

Ostrom has concentrated on how groups of people share the management of common property, such as groups of farmers who have to share water for irrigation, or fishermen who have to select locations for setting out their nets, a set of conditions usually called the tragedy of the commons. The condition of shared access to common resources is a tragedy because selfish actors can exhaust the resource they have access to, as with shepherds overgrazing sheep on a common green or farmers overirrigating from a shared source of water. Neoclassicial economics assumes that to prevent this outcome, a market has to be established where these resources can be privatized and then bought and sold, or that a state agency has to regulate the management of the common property. Ostrom has shown that in some cases the group using the resource can manage it better than either the market or the state. Such arrangements among the group often rely on repeated communications and interactions among the participants, usually in a physical spot common to the participants. Ostrom's work noted that such shared management often relied on mutually visible action among the participants, credible commitment to the shared goals, and group members' ability to punish infractions. When these conditions are met, people with the largest stake in the resources can do a better job both in managing the resource and in policing infractions than can markets or government systems designed to accomplish the same goals.

Groups that manage common resource problems assume a shared commitment to a norm of cooperation. This is different from the ability to see bad behavior and punish it. The easiest infraction to deal with is the one that doesn't happen, so having members internalize a sense of right and wrong when dealing with irrigation or fishing rights becomes an essential tool. This internalization relies on the finding demonstrated by the Ultimatum Game; namely that people in social circumstances will moderate their behavior to be less selfish.

The social reduction of selfish impulses can be triggered easily. When a plate of doughnuts is set out in a common area, office workers will take fewer if there are paper cutouts of eyes nearby (thereby proving H. L. Mencken's hypothesis, "Conscience is the little voice that tells you someone might be looking"). Similarly, the Copenhagen airport in Denmark has taken to posting its rules via cardboard cutouts of airport employees, photographed holding up signs like "Don't take luggage carts up the escalator." The informational content of the sign is no different from before, but having it held up by a picture of a person triggers the sense that human beings are behind all the rules and requests.

Under the right circumstances, we are good at coordinating our actions with regard for other people, even those not present. This skill isn't universal, however; it requires figuring out how to encourage mutual regard for one another and balance selfish motives against it. That challenge is part of any group dynamic—the Z-Boys and the Impressionists had both competitive and collaborative aspects. What's new is the prospect of creating that mutual regard across much larger and more widely dispersed groups, groups who pool their efforts without sharing a physical location, and whose creations can be valuable not just for the participants but for the rest of the world as well.

SOCIAL PRODUCTION: PEOPLE YOU DON'T KNOW, MAKING YOUR LIFE BETTER FOR FREE

If you've used the web twice this week, chances are you owe Brian Behlendorf a thank-you note. Behlendorf was the primary programmer for Apache, the world's most widely used webserver, the software that delivers webpages to your computer. Webservers are easily one of the most important pieces of software ever developed; with something like two billion people now using the web, we make trillions of requests for webpages each year, most of them from Apache servers.

Webservers come in many varieties, but Apache is far and away the most widely used. It's been around for a decade, and for most of that time it has had better than a 60 percent market share. Now, this is nice for the people working on Apache, but so what? Procter and Gamble has a large market share as well, but no one owes their executives thank-you notes—they have the market to reward them. Apache, though, is different, because it's free—as in free speech and free beer (as free software advocates like to say).

The computer code that constitutes Apache is as readily and publicly available as Harry Potter fan fiction, but it is considerably more valuable, and its value is secured by the Apache license, a form of copyright that guarantees that no one, not even its creators, can prevent versions of Apache from continuing to circulate in freely available forms. You can take the code, make your own version of Apache, and sell it, but you can't prevent anyone else from making a competitive version that they can give away for free. The practical effect of this license (and the reason Behlendorf and his colleagues developed the license and not just the software) is to ensure that anyone making an improvement to Apache can share it easily, without fear of later being alienated from their work. The license makes access to Apache a right for its programmers and a gift to its users.

The Apache project unites an incredible range of different talents; some contributors are trying to break new ground, some are trying to make the current version run faster, and some are simply trying to fix bugs. No one person has the necessary talents to do all that work alone, but groups of people can be destabilized by the conflicting desires of the individuals involved. The legally enforced equality of access and freedom for unlimited use of Apache means that while people can (and do) make commercial versions of the code, most of the programmers working on it will work on the free version. Furthermore, because anyone can modify a version of Apache for his or her own private use, the license encourages a huge amount of experimentation, and the results of those experiments can end up being reintegrated into the main version. Low hurdles to participation make both research and the incorporation of results easier than for a commercially developed product.

The advances in Apache (and in all large free software projects) rely on the existence of a collaborating group of people, and the ability to recruit that group and integrate their work has driven Apache's decade-long dominance. Apache doesn't just happen to be noncommercial; it has to be noncommercial in order to be able to take in contributions from as many people as it can as cheaply as it can. Restricting access to paid employees or restricting use to paying customers would set up obstacles that would prevent Apache from being as robust, flexible, and popular as it is now. People do get paid to work on it; IBM pays hundreds of engineers to work on various open source projects like Apache, which results in a more valuable product, just as IBM pays engineers to work on software that IBM owns. In the case of Apache, though, they can't manage or control the project just because they pay some of the workers. The payment to the engineers creates value, but that value doesn't come with either managerial or property rights.

Free software projects instead rely on Elinor Ostrom's mechanisms of joint governance of commonly accessible resources, with lots of communication among the members, repeated interactions, and a mutually binding agreement (like the Apache license). Both the community and the license are crucial to Apache's success; the community polices the license's provisions, and the license provides the community with a measure for acceptable behavior. The project would fail without either of these components.

Apache's method of organizing group activity is both old and new. Like other collaborative circles, it started with a core group of programmers, a team of half a dozen people in San Francisco, gathered around Behlendorf but not managed by him, who were mainly working to improve the software they used. As the usefulness of the software grew, the group of contributors grew as well, and as the contributors grew, the utility of the software grew. The group of contributors expanded to include dozens, then hundreds, and now thousands; however, they didn't all become members of the core group. Instead, they became peripheral members, whose individual efforts were usually less significant than those of the founders, but whose aggregate contributions were enormously important for improving Apache—millions of small additions and fixes adding up to continuous positive change—and for spreading its use far and wide. For that group of contributors, Apache offered the same kind of coordinating value as an empty swimming pool or a Parisian painting studio.

Unlike older models of collaborative circles, however, the Apache effort is global—the project has taken in additions and improvements from programmers in dozens of countries. It is also largely virtual, balancing occasional face-to-face meetings among participants with lots of online work and conversation. And its mechanism of guaranteeing that participants can always access their work, the Apache license, is a bit of legal structure created precisely to support this global and virtual style of collaboration. The Apache project demonstrates that we can now create group efforts that operate on a large scale, without taking on all the costs usually associated with groups that large.

When we want something to happen, and it's more complex than one person can accomplish alone, we need a group to do it. There are many ways to get groups to undertake big or complex activities, but for large-scale, long-lived tasks, the primary mechanisms have been twofold. The first is the private sector, where a task will get done when the group to do it can be assembled and paid for less than their output will fetch in the market. (This is the world of the firm; it is how most cars are built.) The second is the public sector, where employment comes with an obligation to work together on tasks that are of high perceived value, even if they are not compensated in the market. (This is the world of government and nonprofits; it is how most roads are built.) The single most heated political debate in the last century was how best to balance the competing values of those two modes. The result, after the collapse of Communism as the maximum case for a pure public option and after the rise of the welfare state tempered the idea of a pure market, has been a convergence to a broad center, with different mixes of public and private creation in different places.

There is a third mechanism for group production though, outside managed organizations and the market. Social production is the creation of value by a group for its members, using neither price signals nor managerial oversight to coordinate participants' efforts. (This is the world of friends and family; it is how most picnics happen.) Social production was not included in the heated political debates of the twentieth century, because the things people could produce for one another using their free time and working without markets or managers were limited.

Two things happened to end that consensus. First, behavioral economics upended the idea that humans always determine value rationally, the way competitive markets do. In fact, we aren't rational, we are "predictably irrational" (to use the title of Dan Ariely's wonderful 2008 book on behavioral economics), and markets turn out to be a special case, effective only under tightly controlled conditions. As with the Ultimatum Game, the default human behavior relies on mutual regard for other participants, even when there's money to be made. The second thing that has happened is that the emergence of a medium that makes group coordination cheap and widespread caused many of the old limits on social production to recede.

This is the mechanism of production that Harvard law professor Yochai Benkler has called "commons-based peer production," work that is jointly owned or accessed by its participants, and created by people operating as peers, without a managerial hierarchy. The inclusion of millions of new participants in our media environment has expanded the scale and scope of such production dramatically. Where markets and managers have been the preeminent mechanisms for large-scale creation, we can now add this form of social production as a way to take on such tasks, linking our aggregate free time to tasks we find interesting, important, or urgent, using media that now provides opportunities for this kind of production. This increase in our ability to create things together, to pool our free time and particular talents into something useful, is one of the great new opportunities of the age, one that changes the behaviors of people who take advantage of it.

GENERATION X, Y, Z

Our theory-induced blindness around human motivation can keep us from reexamining beliefs about why people behave as they do. You could see how common beliefs got in the way of understanding new behaviors early in the last decade, with the appearance and rapid growth of a piece of music-sharing software called Napster.

Invented in 1999 by Shawn Fanning, a nineteen-year-old computer science student in Boston, Napster let users share music with one another. Its mechanism was simple: Napster users could share a list of the songs they had on their computer (so long as those songs were saved in the MP3 audio format). This list was combined with lists from other Napster users, creating a master list for all the music held by Napster users worldwide. Then if you decided you had to have a copy of Vanilla Ice's "Ice Ice Baby," Napster could tell you which other users had it. Once you had this information, you could then get a copy directly from that user's computer.

It worked like a Yellow Pages for music. Just as you can open up a phone directory to find a plumber and then call the plumber directly to arrange an appointment, people would look up the location of music on Napster's master list, then copy the music directly from other users. Napster users could do it all for free, because making a perfect copy of a song (or of anything stored on a computer) is a side effect of owning the computer itself.

Napster acquired tens of millions of users in less than two years, making it the fastest-growing piece of software of its day. Its astounding success surely said something about the culture, and two conflicting interpretations were advanced in the early 2000s. The first was that young people had all become morally corrupt, willing to flout the sacred conventions of intellectual property. The second was that young people were so imbued with the spirit of sharing that they were happy to engage in the communal opportunity that Napster offered. The first explanation purported to explain why young people were so willing to take, the other why they were so willing to give. Both explanations couldn't possibly be correct. In fact, neither of them was correct.

One of the weakest notions in the entire pop culture canon is that of innate generational difference, the idea that today's thirtysomethings are members of a class of people called Generation X while twentysomethings are part of Generation Y, and that both differ innately from each other and from the baby boomers. The conceptual appeal of these labels is enormous, but the idea's explanatory value is almost worthless, a kind of astrology for decades instead of months.

Generations do differ, but less because people differ than because opportunities do. Human nature changes slowly but includes an incredible range of mechanisms for adapting to our surroundings. Young people born in the decades after the baby boom ended were labeled Generation X, and they began entering the workforce in real numbers in the late 1980s. Gen Xers were said to be lazy—"slackers" in the parlance of the time—who didn't exhibit the straightforward work ethic of their predecessors. (As someone born at the tail end of the baby boom, I loved this reasoning.) Commentators wrung their hands about the slackers in our midst, further evidence that society was going to hell in a handbasket. (Remember the Gin Laws?)

Then in the early 1990s a funny thing happened: Gen Xers

started founding companies, joining start-ups, and working around the clock in pursuit of new opportunities. Gen Xers weren't slackers at all—they were entrepreneurial! How could we have gotten it so wrong?

It's simple: we didn't factor in the environment in which the then-twentysomethings were living. The market crash of 1987 was followed by a fitful performance in the U.S. economy that, by the early 1990s, had tipped into a full-blown recession. In a recession, taking a dead-end job and conserving costs by hanging out with friends and drinking cheap beer are perfectly sensible responses. Maybe this generation wanted to be go-getters even in the depths of the recession, but people don't behave in ways they don't have the opportunity to behave in. Once the recession was over, the landscape of opportunity changed dramatically: it became easier to find a well-paying job, to start a company, or to join a start-up, all activities that the former slackers dove into with gusto.

At the moment of their earliest adulthood, Gen Xers were entering an economy that was inimical to ambition, and they behaved accordingly. Then, fairly suddenly, the economy started rewarding ambition, and the supposedly core psychological attributes of those young people simply vanished, to be replaced by an almost opposite set of attributes. You'd think this transformation would have broken people of their faith in such generalizations, but the desire to attribute people's behavior to innate character rather than to local context runs deep. It runs so deep, in fact, that psychologists have a name for it: the fundamental attribution error. The fundamental attribution error is at work when we explain our own behavior in terms of the constraints on us ("I didn't stop to help the stranded driver because I was late for work") but attribute the same behavior in others to their character ("He didn't stop to help the stranded driver because he's selfish"). Similarly, we fell into the fundamental attribution error when we thought Gen Xers weren't working hard because they were lazy.

Theories of generational difference make sense if they are expressed as theories of environmental difference rather than of psychological difference. People, especially young people, will respond to incentives because they have much to gain and little to lose from experimentation. To understand why people are spending so much time and energy exploring new forms of connection, you have to overcome the fundamental attribution error and extend to other people the set of explanations that you use to describe your own behavior: you respond to new opportunities, and so does everybody else, and these changes feed on one another, amplifying some kinds of behavior and damping others. People in my generation and older often *tut-tut* about young people's disclosing so much of their lives on social networks like Facebook, contrasting that behavior with our own relative virtue in that regard: "You exhibitionists! We didn't behave like that when we were your age!" This comparison conveniently ignores the fact that we didn't behave that way because no one offered us the opportunity (and from what I remember of my twenties, I think we would have happily behaved that way if we'd had the chance).

The generational explanations of Napster's success fall apart because of the fundamental attribution error. The recording industry made that error when it became convinced that young people were willing to share because their generation was morally inferior (a complaint with obvious conceptual appeal to the elders). This thesis never made sense. If young people had become generally lawless, we'd expect to see a rise not just in sharing music but also in shoplifting and other forms of theft. Instead, the recording industry was bemoaning the rising criminality of youth in a period characterized by a reduction in crime almost everywhere in the industrialized world. Even crimes against property were falling. It looked as if ever-more-law-abiding youth were engaging in a special form of criminality, one that applied only to digital data, because of its unique characteristics.

Before the rise of file sharing, you could give a CD to your friend, but once they had it, you didn't. This was the normal case for recorded music, indeed for any physical object. Sharing a book or a magazine or a pair of shoes is what economists call "rival" sharing; if I have my copy of Vanilla Ice's *To the Extreme*, you don't, and if you have it, I don't. A song on your computer, though, is different, because with digital music I can give you a copy while keeping mine. I can sensibly refuse to lend you my Vanilla Ice CD if I still want to listen to it, but refusing to let you copy Vanilla Ice songs off my computer would be different, since copying costs me nothing and inconveniences me not at all. As Thomas Jefferson famously remarked: "He who receives ideas from me, receives instruction himself without lessening mine, as he who lights his taper at mine receives light without darkening me." Napster, like all forms of digital data sharing, took advantage of the fact that music could now be shared like thoughts rather than like objects.

People who hailed Napster as evidence of a communitarian generation were also making a fundamental attribution error, mistaking a new behavior for a change in human nature rather than a change in opportunity. Young people using Napster weren't inherently more community-minded; they simply wanted music for free. Without that desire, Napster would have failed.

The decision not to make someone else's life better when it would cost you little or nothing has a name: spite. The music industry, in order to preserve its revenues, wanted (and still wants) all of us to be voluntarily spiteful to our friends. Fanning designed a system that shaped the users' cumulative behavior away from spitefulness and toward sharing; and like all applications that rely on the cumulative participation of the users, Napster provided the means to share, but only the users could create the actual value for one another. As the visionary Kevin Kelly wrote in an essay called "Triumph of the Default," engineers can influence the behavior of their users:

> Therefore the privilege of establishing what value the default is set at is an act of power and influence. Defaults are a tool not only for individuals to tame choices, but for systems designers—those who set the presets—to steer the system. The architecture of these choices can profoundly shape the culture of that system's use.

Kelly's conclusion that defaults allow the designer to steer the system is critical. Defaults don't drive the system, because they don't create the motivations to use it. They simply steer those motivations to certain outcomes, provided the users are interested. Fanning designed Napster so that the default behavior was communal sharing. Two things were required to pull this off. The first was a medium that made sharing vanishingly cheap, and the second was a system of defaults that encouraged sharing.

Napster spread among the young not because they were more criminally minded than their elders, nor because they were possessed of a greater spirit of sharing. Napster spread for three much more prosaic reasons: (1) digital data is infinitely and perfectly copyable at zero marginal cost; (2) people will share if sharing is simple enough, and we generally resist being spiteful under the same conditions; and (3) Shawn Fanning designed a system to link (1) to (2) via the right incentives. That's it. That's what turned the recording industry upside down. Similarly, Napster's original model was destroyed when the recording industry's legal actions raised the cost of sharing high enough to unlink (1) and (2) for a significant number of people.

If that explanation sounds boring, well, it is boring, especially when compared to tales about how society is (pick one) going to hell in a handbasket or entering a period of higher consciousness. The rise of music sharing isn't a social calamity involving general lawlessness; nor is it the dawn of a new age of human kindness. It's just new opportunities linked to old motives via the right incentives. When you get that right, you can change the way people interact with one another in fairly fundamental ways, and you can shape people's behavior around things as simple as sharing music and as complex as civic engagement.

COLLABORATIVE SPIRALS

In Lahore, Pakistan, Sabrina Tavernise of the *New York Times* observed a trio of young men who were tired of their country's divisive politics and of its weak government that was unable to provide even basic services. Inspired by Pakistani lawyers who had turned out in the streets earlier in the decade to protest government interference on Pakistan's High Court, the three men, Murtaza

Kumail Khwaja, Saif Hameed, and Omar Rasheed, decided to mobilize people in the street to pick up garbage. Garbage is a classic negative externality, a bad by-product of someone else's actions. Someone who drops trash on the street is glad to be rid of it; the negative effects of that action, however, are cumulative, and mostly impact other people. Negative externalities require collective action to reverse. Sometimes that collective action is funded by taxes, other times by elbow grease.

The problem with the elbow grease model is that the cost of coordinating friends and neighbors is often prohibitively high. The Lahori youths, calling themselves the Responsible Citizens, solved this problem by using Facebook to recruit their friends, since Facebook lowers the cost of social coordination among its users. Once the three had enough recruits to join them, they showed up on the streets on Sundays, gathering trash from a public market in Anarkali. Local citizens and merchants at first merely observed, but as the Responsible Citizens kept returning, the locals began to join them. This new labor, in turn, helped the Citizens expand to other markets in town. Sociologists call this kind of behavior positive deviance. Any community has members who deviate from social norms in negative ways, engaging in antisocial behavior or even criminality even when they have opportunities and resources similar to those of the other members. Positive deviants are those who behave better than the norm, even when faced with similar limitations or challenges.

Khwaja, Hameed, Rasheed, and their recruits were deviating positively from Lahore's norm of civic passivity. The immediate effect of their actions was to reduce the amount of trash on a few market streets, but their longer-term value is not their output but their example. As they put it in a Responsible Citizens manifesto: "We wish to nurture in everyone a community spirit." They were trying to make civic action contagious.

This idea is less crazy than it sounds. In 1973 Mark Granovetter showed in a seminal paper, "The Strength of Weak Ties," that people tend to find jobs through casual acquaintances rather than through close friends or family. Since then an increasing body of research has demonstrated the importance of social networks to our well-being. Nicholas Christakis and James Fowler, researchers at Harvard Medical School, have shown that social networks spread all kinds of behaviors: we are likelier to be obese if our friends are obese, or to exercise if they exercise, or even to be happy if they are happy. More remarkably, we are susceptible to the traits of even members of our social networks beyond our immediate acquaintances. The more the friends of our friends are happy, the likelier we are to be happy, and even if our friends' friends' friends are happy. Habits and traits spread through social networks through up to three degrees of separation, and though these traits are not contagious like a virus, they are contagious in that they spread through social contact.

Seen in this light, the Responsible Citizens are not just cleaning up trash—they are attempting to demonstrate positive civic engagement to the people they know, and to the people those people know, and to the people *those* people know. It's too soon to handicap the long-term effects of the Responsible Citizens, but without social contagion, their task would be hopeless. With it, their work may help create social change even among strangers. We create one another's opportunities, whether for passivity or for activity, and we have always done so. The difference today is that the internet is an opportunity machine, a way for small groups to create new opportunities, at lower cost and with less hassle than ever before, and to advertise those opportunities to the largest set of potential participants in history.

Twentieth-century beliefs about who could produce and consume public messages, about who could coordinate group action and how, and about the inherent and fundamental link between intrinsic motivations and private actions, all turned out to be nothing more than long-term accidents. Those accidents are being undone by new opportunities, created by us, for one another, using abilities afforded us by our new tools. The driving force behind the Responsible Citizens (and the Apache webserver, and the Grobanites for Charity) is the ability of loosely coordinated groups with a shared culture to perform tasks more effectively than individuals, more effectively than markets using price signals, and more effectively than governments using managerial direction.

Social production is not a panacea; it is just an alternative. Although we are better off being able to use it when it is valuable, it brings its own challenges, just as production via firms and governments does. Even the simplest pooled effort or voluntary participation can be fraught with tension among the individual participants, and between those individuals and the group. Like many aspects of social life, this problem has no solution; the dilemma can be addressed only by various compromises, none of them wholly satisfactory. One way to help a group of participants improve their ability to function together is the creation and maintenance of shared culture.

5

Culture

In January 2000, an unusual paper called "A Fine Is a Price" appeared in the *Journal of Legal Studies*. Written by Uri Gneezy and Aldo Rustichini, it was about psychology, though it appeared in a legal studies journal; it was short, in a field given to writing by the yard; it was written in plain (and quite vivid) English; and it attacked a central tenet of legal theory, namely that deterrence is a simple and reliable way to affect people's behavior.

Gneezy and Rustichini described the mainstream theory of deterrence this way: "When negative consequences are imposed on a behavior, they will produce a reduction of that particular response. When those negative consequences are removed, the behavior that has been discontinued will typically tend to reappear." The theory is simple, straightforward, and commonsensical, but as the researchers noted, it was largely untested. They set out to correct that fact in 1998, working with day-care centers in the Israeli city of Haifa as experimental sites.

Day-care is day-care the world over; working parents with children under school age need someone to watch their children during the day. Sometimes day care is set up as a public service, other times as a business, but in either case, the parents and the day-care workers have a potential daily clash of interests: pick-up time. The workers have outside lives, so they want all the kids safely reunited with their parents by a set time. The parents, on the other hand, busy at work or running errands and never entirely in control of their travel time, want some slack to pick up their children later than the appointed hour.

The study's ten day-care centers in Haifa ran until four P.M., though no penalty for picking up children late was specified. Gneezy and Rustichini observed closing time in the centers to see how often parents were late; in a normal week, there were seven or eight late pickups at each center. Then they instituted a penalty at six of the centers: henceforth, they announced, parents would be fined for picking their children up more than ten minutes late, a fine that would be automatically added to their bill. (The other four centers, the control group, operated unchanged to ensure that any observed effects in the six selected schools were the result of the fine.)

The new rule was imposed at the six centers the following week, and its effect on the parents' behavior was immediate: their lateness *increased*. In the first week, the average number of late pickups rose to eleven; to fourteen in the week after that; and to seventeen the week after that. The episodes of lateness finally topped out a month into the experiment at around twenty a week—nearly triple the pre-fine number. Thereafter, for as long as the fine was in place, the number fluctuated, but it never fell below fourteen and remained closer to twenty most weeks. Meanwhile, the number of late pickups in the four control centers didn't change.

From the point of view of deterrence theory, this result was perverse. The fine was small, just ten shekels (about three dollars), but it should still have had some deterring effect; however bad a late pickup was before the fine was instituted, it should have been ten shekels worse after. And even if it was too small to have a deterring effect, it shouldn't have increased the frequency of lateness. And yet that's just what it did.

The pre-fine bargain between parents and teachers was what Gneezy and Rustichini labeled an "incomplete contract"—a set of relations that took place partly in the market but left considerable room for the interpretation of certain behavioral norms, including those around pick-up time. As they noted in their paper, "Parents could form any belief on the matter, as they probably did, and act accordingly." Once the fine was instituted, however, that ambiguity collapsed, along with the behavioral norms that had been established. The fine turned day care from a shared enterprise into a simple fee-for-service transaction, allowing the parents to regard the workers' time as a commodity, and a cheap one at that. The parents assumed that the fine represented the full price of the inconvenience they were causing, and it seemed to remove any fears that they might suffer some unspecified consequence for abusing the workers' goodwill.

Gneezy and Rustichini kept the fine in place for three months, then ended it. Once the fine stopped, however, the number of late pickups per week didn't return to pre-fine levels; in fact, it remained as high as it had been when the fine was in place. Inducing parents to see the day-care workers as participants in a market transaction, rather than as people whose needs had to be respected, had altered the parents' perceptions of the workers, an alteration that outlived the fine itself. One might impose a fine significant enough to deter lateness, the paper noted, but the experiment showed that market transactions are not merely additive to other human motivations; they alter them by their mere presence.

If this finding sounds familiar, that's because it's the same thing Edward Deci noted in 1970 with his Soma puzzle study: both experiments demonstrated that putting a price on something previously outside of market logic can change it fundamentally. To be sure, Deci had offered a positive price (a payment), while Gneezy and Rustichini instituted a negative price (a fine). But as Gneezy and Rustichini noted, a fine is a price. The more important difference between the Soma experiment and the day-care experiment, though, is that the object of the latter experiment wasn't a wooden puzzle; it was living, breathing people—the workers. "A Fine Is a Price" shows that dealing with one another in a market can fundamentally alter our relationships with one another.

Deci's work focused on the personal motivations of autonomy and competence, demonstrating that pricing them reduced them as motivators. Gneezy and Rustichini focused on social motivations, demonstrating that adding a price to a previously nonmarket transaction can reduce our willingness to treat each other as people we might have long-term relationships with. (It echoes the old observation about prostitution, namely, that men are not only paying for sex, they are also paying the women to go away afterward.) The introduction of the fine was not just personal, affecting the behavior of individual parents. It was also social, introducing a new set of relations between the parents and the workers. Culture isn't just an agglomeration of individual behaviors; it is a collectively held set of norms and behaviors within a group. In the case of the day-care centers, introducing the fine killed the previous culture by altering the way the parents viewed the workers, and that culture stayed killed even after the fine was removed.

How we treat one another matters, and not just in a "it's nice to be nice" kind of way: our behavior contributes to an environment that encourages some opportunities and hinders others. In the culture of the Haifa day-care centers, one simple change had a huge effect. Under the previous "incomplete contract," parents and workers had negotiated an informal but acceptable bargain. When that culture came to include an explicit fine, the parents could view the workers as a means to an end, rather than as partners with a mix of social and commercial bonds.

People's behavior toward one another isn't fully described by the market, because market transactions cover only a small part of the repertoire of human behavior. I ran into an example some years ago at the San Francisco airport. I'd previously called the airline to change the return date of my flight and had been told I'd have to pay an additional twenty-five dollars when I checked in at the airport.

When I got to the ticket counter on the day of the flight, I asked the agent for my new ticket and said, "Oh, and there's a twenty-five dollar charge" as I started to dig out my wallet. "No," she replied, "you have to pay a penalty." Thinking I was about to be hit with another fee on top of the one I already knew about, I said, "I was told I'd only owe twenty-five dollars to change the ticket."

"The twenty-five dollars is a penalty," she replied. At that point I realized what was going on. We agreed that I owed the airline twenty-five dollars, but in my mind that was a reasonable fee for the additional work. In her mind, though, it was a punishment for changing my ticket. Further, she was clearly in no mood to hand over the ticket until I acknowledged that. Given that I had a plane to catch, my desire to argue about the matter evaporated, and I handed over what had suddenly become a twenty-five-dollar penalty, took my boarding pass, and flew home.

That ticket agent was in the same position as the Haifa day-care workers; she was part of a transaction that I was regarding as bloodless, and that fact bothered her. Unlike the day-care workers, though, she was able to insist on the emotional component of the transaction before it could be completed. Market-based transactions are one kind of cultural norm that can form in exchanges between human beings, but many others are possible as well. Elinor Ostrom, the economist whose work on public resources appeared in the last chapter, notes that much twentieth-century economics mistakenly assumed that market transactions are an ideal and even default model for human interactions. But some kinds of value can't be created by markets, only by a set of shared and mutually coordinating assumptions, which is to say by culture.

CULTURE AS A COORDINATING TOOL

In 1645, a group of people living in London decided that they would refuse to believe things that weren't demonstrably true. This is tough; we humans have never been terribly good at subjecting our own beliefs to the kind of withering scrutiny that might disprove them. We do have a related skill, however; we are quite good at subjecting other people's beliefs to such scrutiny, and this asymmetry gave them an opening. They committed themselves to acquiring knowledge through experimental means and to subjecting one another's findings to the kind of scrutiny necessary to root out error. This group, which included the "natural philosophers" (scientists) Robert Boyle and Robert Hooke and the architect Christopher Wren, was referred to, in some of Boyle's letters, as "our philosophical college" or "our Invisible College."

The Invisible College was invisible relative to Oxford and Cambridge, because the members had no permanent location; they held themselves together as a group via letters and meetings in London and later in Oxford. It was a college because their relations were collegial—they operated with a sense of mutual interest in, and respect for, one another's work. In their conversations, they would outline their research according to agreed-upon norms of clarity and transparency. Robert Boyle, one of the group's members and sometimes called the father of modern chemistry, helped establish many of the norms underpinning the scientific method, especially how experiments were to be conducted. (The motto of the group was *Nullis in Verba*—"Believe nothing from mere words.") When one of their number announced the result of an experiment, the others wanted to know not just what that result was but how the experiment had been conducted, so that the claims could be tested elsewhere. Philosophers of science call this condition falsifiability. Claims that lacked falsifiability were to be regarded with great skepticism.

Within a few years, several members of the Invisible College had produced advances in chemistry, biology, astronomy, and optics, and they had developed or improved several key experimental tools, including pneumatic pumps, microscopes, and telescopes. Their insistence on clarity of method made their work both collaborative and competitive, and new methods and insights quickly became input for still further work.

Much of the members' practical work involved chemistry. They were strongly critical of the alchemists, their intellectual forebears, who for centuries had made only fitful progress. By contrast, the Invisible College put chemistry on a sound footing in a matter of a couple of decades, one of the most important intellectual transitions in the history of science. What did the Invisible College have that the alchemists didn't? It wasn't their tools—chemists and alchemists both started out with vials, braziers, and scales. Nor was it insight—no single figure suddenly advanced chemistry, as Newton did with physics. The Invisible College had one big advantage over the alchemists: they had one another.

The problem with alchemy wasn't that the alchemists had failed to turn lead into gold—no one could do that. The problem, rather, was that the alchemists had failed uninformatively. As a group, the alchemists were notably reclusive; they typically worked alone, they were secretive about their methods and their results, and they rarely accompanied claims of insight or success with anything that we'd recognize today as documentation, let alone evidence. Alchemical methods were hoarded rather than shared, passed down from master to apprentice, and when the alchemists did describe their experiments, the descriptions were both incomplete and vague. As Boyle himself complained of the alchemists' publications, "Hermetic Books have such involved Obscuritys that they may justly be compared to Riddles written in Cyphers. For after a Man has surmounted the difficulty of decyphering the Words & Terms, he finds a new & greater difficulty to discover the meaning of the seemingly plain Expression."

This was hardly a recipe for success; even worse, no two

people working with alchemical descriptions could reliably even fail in the same way. As a result, alchemical conclusions accumulated only slowly, with no steady improvement in utility. Absent transparent methods and a formal way of rooting out errors, erroneous beliefs were as likely as correct ones to be preserved over generations. In contrast, members of the Invisible College described their methods, assumptions, and results to one another, so that all might benefit from both successes and failures. The Invisible College became so important to British science that its members formed the core of the Royal Society, a much less invisible organization chartered in 1662 and still in operation to this day.

Culture—not tools or insights—animated the Invisible College and transmuted alchemy into chemistry. The members accumulated facts more quickly, and were able to combine existing facts into new experiments and new insights. By insisting on accuracy and transparency, and by sharing their assumptions and working methods with one another, the collegians had access to the group's collective knowledge and constituted a collaborative circle. Their cultural norms transformed the alchemists' slow accumulation of personal and idiosyncratic beliefs into a set of methods and results that could be observed, understood, and recombined by any scientifically literate participant.

Combinability makes knowing something different from having something. If you have a stick, and someone gives you another one, you have two sticks. It's better than having just one, but it's still not much. If, on the other hand, you have a piece of knowledge—that rubbing the two sticks together in a certain way can make fire—you can do something of value that you couldn't do before.

Increasing the number of things you have can be useful, but increasing the amount of knowledge you have can be transformative. This is what makes the ways a society shares knowledge so critical, and what helped give the Invisible College such a dramatic edge over the alchemists. Even when working with the same tools, they were working in a far different, and better, culture of communication.

THE ECONOMICS OF SHARING

Knowledge is the most combinable thing we humans have, but taking advantage of it requires special conditions. In his book *The Economics of Knowledge,* Dominique Foray, a French economist at the École Polytechnique Fédérale in Switzerland, identifies these conditions as the size of the community, the cost of sharing that knowledge, the clarity of what gets shared, and the cultural norms of the recipients.

The size of the community, the first condition, is fairly intuitive. Knowledge, unlike information, is a human characteristic; there can be information no one knows, but there can't be knowledge no one knows. A particular bit of knowledge lives only in minds capable of understanding it. The community that can understand the lyrics to "Happy Birthday" is much larger than the community that can understand Sanskrit poetry. (Literacy is critical, because it increases the size of the community that can make use of any given bit of knowledge.) The more people in a community who can understand a particular fact, method, or story, the likelier it is that those people will be able to work together to make use of those bits of knowledge.

The second condition that affects combinability is the cost of sharing knowledge. Anything that lowers the cost of transmitting knowledge can increase the pool of knowers. When the printing press lowered the cost of both making and owning books, it hugely increased the number of people who could read any given book, and the number of books a literate citizen could read in a lifetime. The spread of the telegraph famously brought international news to many local newspapers, a fact that often occasioned complaint—a local paper in Michigan canceled its telegraph service because there was too much global news and "not a line about the Muskegon fire"—but the lowered cost of knowing things from around the globe affected not only what people knew but how they behaved. The first great wave of modern globalization was driven in part by the telegraph's lowering of the costs of sharing information. Today the internet is lowering the cost of transmitting not only words but also images, video, voice, raw data, and everything else that can be digitized, a change in cost on a par with that of the telegraph and the printing press.

Foray's third condition for combinability is clarity of the knowledge shared. We communicate instructions about cooking in recipe form for a reason: by listing ingredients and ordering instructions in steps, a recipe is clearer than a purely narrative description of how to cook a dish. A rambling description might have the same informational content as a recipe, but the form of a recipe is clearer. As a result, once any field of endeavor acquires something like a recipe—a set of instructions for an activity, separable from the activity itself—it can circulate much more effectively among people who can understand it.

The spread of recipelike clarity can accelerate the sharing of knowledge among groups working on the same problem, but it can also make it easier for others to benefit from the knowledge so produced, because clear expression of an idea can travel from person to person and group to group more easily than the same idea expressed in a way that only the members of a specific group can understand. (This principle echoes Boyle's critique of the "involved Obscuritys" in alchemical books.) Eric von Hippel, the scholar of user-driven innovation quoted in Chapter 4, studied a kite-sailing community called Zeroprestige, which designed kites using 3D rendering software. After producing several such designs and posting them online, the members of Zeroprestige were contacted by a manufacturer in China who offered to make those designs into functioning sailing kites. When a factory owner goes looking for design talent instead of vice versa, the logic of outsourcing is turned on its head; it was possible only because the description of the kites, which was written in a standard format for 3D software, was enough like a recipe for the manufacturer to be able to discover them online and to interpret them without help.

Increases in community size, decreases in cost of sharing, and increases in clarity all make knowledge more combinable, and in groups where these characteristics grow, combinability will grow. These three conditions are all magnified by a medium that is global and cheap, and that lets unlimited perfect copies of information spread at will, even among large and physically dispersed groups. Our technological tools for making information globally available and discoverable, by amateurs, at zero marginal cost, thus represent an enormous and positive shock to the combinability of knowledge.

These three conditions—community, cost, and clarity—aren't enough, however, as we know from the Invisible College. Foray's fourth condition is culture, a community's set of shared assumptions about how it should go about its work, and about its members' relations with one another. To really take advantage of combinability, in other words, a group has to do more than understand the things its members care about. Its members also have to understand each other, in order to share or work together well. Etienne Wenger, an independent sociologist, coined the term "communities of practice" to describe people who come together to share their knowledge as a way of getting better at what they do. Communities of practice, says Wenger, exchange and interpret information, help their members attain and retain competency at their shared job, and, critically, provide homes for the identities of the practitioners. A community of practice, in other words, is more about maintaining the cultural norms that hold the community together than about preserving any particular bit of knowledge the community holds. Knowledge in such communities often changes, but cultural commitment to the work must remain in place throughout the group's life span.

Our new tools provide an opportunity to create new cultures of sharing, and only in the hands of these cultures will our ability to share become as valuable as it can be. The community of practice that created the Apache webserver (see Chapter 4) operates in an almost completely transparent manner. Not only does that community share the computer code; it shares discussions and arguments about how to improve that code as well. Free software projects are thus home not just to a valuable object—the software—but to a valuable culture: the participants' ways of dealing with one another and of deciding which changes to the code constitute a real improvement. Computer code can be hard to read, and even expert programmers have difficulty telling, just by looking at the code, whether a piece of software will work well or poorly. But in the Apache culture, the projects that are making progress are identifiable by looking at the way the creators of code communicate with one another. Projects that cause constant, passionate argument among programmers are usually healthy, while those that generate little debate or communication are languishing. According to an open source aphorism, "Good community plus bad code makes a good project." If an open source project's community members are committed to improvement, then they will commit to the hard work of identifying and making those improvements. Moreover, a project with a lot of room for simple improvements will be more enticing to that community, at least in the early days. One that has too much wrong will repel community rather than attract it, but if a community in its formative stages can see several places where its work will make a difference, they will be likelier to jump in.

Prior to the 1990s, we had no examples of software that relied on globally distributed volunteers to create a working piece of code. Linus Torvalds, the originator of Linux, is widely credited with this innovation, after he issued a public call for collaborators on the project in 1991, on the global bulletin board system called Usenet. Since that original public invitation, the history of Linux has been as much about keeping the community together and on track as it has been about purely technical issues. Free software works in ways explained by Foray's observation about culture and combinability: the norms of the participants are a critical predictor of success or failure. Those norms are based in part on how the members of the group understand their relationships to one another.

COLLEGE PROFESSORS AND BRAIN SURGEONS

In 2007, Christopher Avenir was a freshman engineering student at Ryerson University, in Toronto. For anyone like Avenir, born in the digital age and college-bound, participation in the media environment is simply a by-product of living in the developed world. By the time he was five, the internet was publicly accessible. By the time he was ten, the web had become a huge success. By the time he was fifteen, social networking sites like Friendster, MySpace, and Facebook had all launched. And before he was twenty, he went to college.

In his first semester at Ryerson, Avenir enrolled in an introductory chemistry class, and like students since time immemorial, he discovered that the work was quite difficult. Seeing this, he decided, again like students since time immemorial, to join a study group. So far, his story is normal, but since he's the age he is in the era he's in, the world provided a new opportunity: online study groups. Avenir joined a group on Facebook called "Dungeons/Mastering Chemistry Solution." (The Dungeon is the affectionate name Ryerson students gave the room on campus where real-world study groups meet.) Avenir became an administrator of the group, responsible for handling requests to join, and 146 of his classmates eventually became members.

Then at the end of the term, Andrew McWilliams, his professor, discovered the group on Facebook, and though he couldn't examine its contents (he wasn't a member), he reported Avenir to the university. Ryerson filed academic charges against Avenir, 147 charges in fact—one for being an administrator of the Facebook group and one for each of his fellow students who joined. This was serious enough to create a threat of expulsion from Ryerson.

The case of Avenir versus Ryerson displays the clash of two deeply held but contesting views that exist within any university. Ryerson's position was simple: the use of Facebook was wrong, because individual work should never be shared. Technology dean James Norrie said, "Are we Luddites here at Ryerson? No, but our academic misconduct code says if work is to be done individually and students collaborate, that's cheating, whether it's by Facebook, fax, or mimeograph." Avenir started from a different position; for him, a Facebook study group was just an extension of the university setting. If study groups met in a place called the Dungeon to work on their chemistry homework, it shouldn't have made any difference whether that place was on campus or online. As Avenir said, "If this is cheating, then so is tutoring and all the mentoring programs the university runs." Even when students are asked to work individually, they learn some of how to do that work from one another. A spectrum runs from peer instruction to collaboration to outright cheating, but when the students gathered together in the Dungeon, the university trusted them to understand and respect the different points on that spectrum. From this point of view, all Facebook provided was a mediated space that served the same role as the Dungeon did.

The different interpretations of Facebook use seem like a simple clash of beliefs—"Facebook is just another medium, like the mimeograph" versus "Facebook is just an extension of the existing social world." The solution should be to pick whichever statement was correct and to implement whichever existing Ryerson policy applied to use of media or to conversations among students. The difficulty, though, is that choosing between those two statements would produce the wrong choice. Facebook is not enough like a fax machine or a mimeograph to be perfectly comparable to older media, because it is more social than older media, and because its participants communicate in groups. (A mimeographed memo does nothing to allow its recipients to talk together.)

Nor is Facebook a simple replacement for a study group in the basement of Ryerson. First of all, Facebook groups are visible worldwide, making them an implicit advertisement for life at Ryerson. Much was made of a request on the Facebook page that read: "If you request to join, please use the forms to discuss/post solutions to the chemistry assignments. Please input your solutions if they are not already posted." Though McWilliams didn't observe the conversation in the group itself, this request seemed to solicit fairly naked answer sharing, contrary to the nature of the assignment. (Students who participated in the group insisted that no such sharing had taken place, though this claim is impossible to judge independently, as the group has been deleted and no public record is available.)

Then there's the question of freeloading. Even though study groups meet in the real-world Dungeon on the Ryerson campus, none of those tables seat 146. If you have a study group of half a dozen people, and a potential new member arrives and says, "I'm not here to participate, I'm here to mooch answers off you guys," that person will be easy to identify and exclude. As Dave Hickey observed about musicians playing in small clubs, it's easy to tell in an intimate setting who's showing up to be part of the event, while in larger groups it's much easier to be a looky-loo, consume without participating.

Online spaces are different. Whatever else the value of a Facebook study group may be, it's a safe bet, in a group of 146, that someone was freeloading, taking advantage of the shared creation of value without offering much in return. Indeed, many online collaborations, whether study groups or open source software or user-created collections of media, are free-rider tolerant, in which small, highly involved groups of people co-create something valuable for a much larger group of people who take advantage of it. Such free-rider-tolerant systems can be tremendously valuable, but they are often a lousy fit for education.

This clash of similes between "Facebook is like a fancy fax machine" and "Facebook is like an online room" tends to overstate the degree to which something new is the same as what went before, and to underestimate differences between old and new. Facebook is in fact a lot like Facebook; Avenir and his classmates used it precisely because it did things that neither fax machines nor tables do. It distributes information among a known social group cheaply, instantly, and without making the participants synchronize themselves in space or time.

These capabilities, as remarkable as they are, are not an obvious positive for education. Ryerson's administration was right to be concerned, because college education has never been solely or even mainly about efficiency, and is less about getting the right answer than learning the right techniques. When an instructor poses a question about chemical combinations of hydrogen and oxygen, for example, it isn't because he doesn't know where water comes from, but because he wants the students to learn how to get to the right answer themselves. Ways of coming up with the right answer that involve simply asking other people, without internalizing the process, don't actually educate the student. In fact, being handed the answer specifically defeats the purpose of education.

Yet communities grappling with technical information share observations and techniques all the time, and, other things being equal, learners who share their observations and frustrations with their peers learn faster and retain more of what they've learned than those who study alone. Round and round it goes, with no one point of view making clear what Ryerson's position should be. It's not at all clear, in fact, that there is a right answer—the rebalancing of individual and group work gets right to the heart of what colleges do, and the usual result is a trade-off with different advantages and disadvantages. The only two points of certainty, in fact, are at the extremes—forbidding anyone to talk to anyone else ever and requiring everyone to talk to everyone else all the time. Neither is useful from an educational standpoint, so some new bargain is imperative.

What is clear is that the simple application of seemingly fundamental principles isn't actually simple, because the principles aren't actually fundamental. Ryerson's policy, and indeed the implicit policies regarding study groups for most colleges and universities, relied on ancient assumptions that hardly needed to be spelled out: Eighteen-year-olds aren't global publishers. Study groups have to meet in real rooms. You can't get 146 people around one table. What happens on campus isn't visible to the entire world. And so on. Ryerson's reaction was driven in part by the sudden need to reorient its policies in an era when those assumptions no longer apply.

At the time of Avenir's disciplinary hearing, much of the public discussion focused on what a raw deal he was getting—he had not founded the Facebook group, but merely become its administrator, so his actions weren't so markedly different from those of his fellow study group members. Furthermore, none of the participants did anything to hide their decision to join—they even named it after the real Dungeon, making it hard to believe that they thought they were cheating. (Consider how easy it would have been to set up a secret mailing list—had they actually wanted to cheat, McWilliams would never have known.) Ryerson seems to have subsequently decided that its initial charges against Avenir were something of an overreaction; he was graded down for the particular test that was taking place at the time of the study group, but he wasn't expelled.

Any attempt to ban the use of social media would have committed Ryerson to a degree of surveillance incompatible with treating students like adults. Instead, limits on the use of social media have to be enforced mainly by the students themselves, both as a matter of personal discipline and as part of their cultural expectations of one another. The community at Ryerson (and, indeed, at all educational institutions) has nothing left but to forge a new bargain, explaining to students which modes of sharing are okay and which aren't. That bargain involves actively determining the balance between individual and group inquiry, which real-world limitations once held in place. Study groups were limited to face-to-face interaction because of the lack of alternatives; with that constraint gone, students have to be involved in forging new constraints in the context of their new capabilities.

Society is shaped as much by inconvenience as by capability, by what it can't do as by what it can. Those two characteristics are deeply imbalanced, however, because the cultural assumptions that rise up around inconvenience simply seem like realism: study groups are small because large, active groups can't work together on short deadlines. Groups of friends and neighbors carpool because there's no way to match supply and demand at larger scale. Professionals have to create reference works because volunteer labor can't be coordinated sufficiently well to make anything of value. And so on. Managing inconvenience, big or small, often involves creating a particular class of worker. College professors, restaurant reviewers, librarians, and file clerks all help keep the inconvenience to manageable levels for everyone else.

When things that used to be inconvenient stop being inconvenient, though, the old accommodations have to be renegotiated, including the role of those workers: when you can get advice about a restaurant from the aggregate view of people who've actually eaten there, the value of the critic as a source of recommendation is reduced. Other functions of the critic, such as interpreting its chef's intentions or relating it to the history of a particular cuisine, remain, but the overall value of the reviewer's work shrinks because the world has changed around him.

This change can be disorienting and has led many critics to assault just such aggregates of popular experience. One early critical complaint was an essay called "The Zagat Effect," written by Steven Shaw in 2000. Zagat is a restaurant guide that aggregates user-generated reviews and ratings. Shaw complained bitterly about them, focusing in particular on their ranking New York City's Union Square Café as number one, which he felt was unjust:

> [Union Square Café] is number one in the sense that it emerges first in response to this question on the survey: "What are your favorite New York restaurants?" . . . Union Square Café is, indeed,

> a very good restaurant, one beloved by many New Yorkers for its compassionate service—it is perhaps the most unintimidating of the city's better restaurants—and its simple but intensely flavorful food. But with all due respect to that justly popular establishment, it is patently ridiculous to rank it ahead of a dozen other places, and in particular such world-class restaurants as Lespinasse, Jean Georges, and Daniel.

Nowhere does Shaw spell out *why* preferring Union Square Café to Lespinasse is patently ridiculous—calling Lespinasse world-class simply begs the question. In a world where access to information is open, the critic does a delicate dance. Shaw is unwilling to condemn Union Square as a bad restaurant; it's just not the kind of restaurant people like him prefer, which is to say people who eat in restaurants professionally and are happy to have a little intimidation with their appetizers. But if he makes that complaint too visibly, he risks undermining his desire to be able to guide his audience. Back when professional reviews were the only publicly available judgment of restaurants, this difference didn't matter much (and critical contempt for the audience wasn't so visible), but when we can all now find an aggregate answer to the question "What is your favorite restaurant?" we want that information, and we may even prefer it to judgments produced by professional critics.

A common objection to the spread of shared knowledge is the need for professional skill, an idea often expressed with the observation that you wouldn't want brain surgery performed by someone who learned their craft from Wikipedia. Let us stipulate, as the lawyers say, that this is true; when surgery on the brain is called for, having it done by an accredited surgeon seems like a good idea. The funny thing about this rule, though, is that we don't really need it, because it is self-evident. The stock figure of the amateur brain surgeon comes up only in conversations that *aren't* about brain surgery. The real assertion is that every time professionals and amateurs differ, we should prefer the professionals, and brain surgery is just one illustrative example.

There are two weaknesses in this line of thought. The first is that you wouldn't want a brain surgeon who learned everything he knew from *Encyclopaedia Britannica* either. The brain surgery analogy isn't broadly applicable, because it says nothing about deciding between competing sources of information. Here's an alternate assertion: you should never eat at a restaurant without being guided by a professional restaurant critic. After all, who knows what could happen? You could end up eating at places with simple but intensely flavorful food, and no intimidating waitstaff anywhere in sight. This example is as ridiculous as the brain surgery one, but at the other extreme. But it offers us a range of analogies, and we can now ask, of any given function, "Is this more like being a brain surgeon or a restaurant reviewer?" Brain surgeons know the parts of the brain, and they also know how to wield a scalpel; slicing up parts of real brains is a job that must be limited to professionals, but it's not clear that knowing the names of the parts of the brain must be similarly limited.

The second weakness in the brain surgeon analogy is that it invites the hearer to assume that we should always go with a professional over an amateur. But curiously, no one believes this proposition, not even the people fretting about Wikipedia-trained brain surgeons. In fact, were this preference for the professional universally applied, we would all be patronizing prostitutes—they are, after all, far more experienced in their craft than most of us will ever be. By comparison, people in love are amateurs (in the most literal meaning of the word). But here intimacy trumps skill. For similar reasons, I sing "Happy Birthday" to my children, even with my terrible singing voice, not because I can do a better job than Placido Domingo or Lyle Lovett, but because those talented gentlemen do not love my children as I do. There are times, in other words, when doing things badly, with and for one another, beats having them done well on our behalf by professionals. Chris Anderson, author of *Free,* tells a story about his kids' deciding what to watch on a big TV monitor. His kids are *Star Wars* fans and were given a choice between seeing one of the *Star Wars* movies in high definition and watching YouTube videos of *Star Wars* scenes reenacted with Legos. You can guess which they chose. They already understood the *Star Wars* canon—the novelty came from seeing what their peers were doing with that shared knowledge.

Two countervailing forces, in other words, pull against a bias toward pure professionalism. The first is the Zagat's counterexample (the value in ordinary people sharing what they know) and the second is the "Happy Birthday" counterexample (the value in doing something that makes you feel a sense of membership or generosity). Sometimes the value of professional work trumps the value of amateur sharing or a feeling of belonging, but at other times people find large-scale and long-lived sharing better. As more people come to expect that amateur participation is always an open option, those expectations can change the culture.

PATIENTS LIKE US

To continue with medical analogies, let's imagine a patient who actually has a complicated and life-altering disease. Needless to say, such a patient wants a professional to do his diagnosis and treatment, but he also wants to know something about the treatment the doctor prescribes. The doctor won't have time to give the patient all the information he wants. Here both kinds of amateur value—sharing and a feeling of belonging—can now come to the fore in ways they couldn't even a few years ago.

PatientsLikeMe.com is a site that, true to its name, allows patients with similar chronic health conditions to share information and offer support. The advantages of joining a group like this fit the transactional model of the current health-care system: patients can learn from one another about how to manage long-term or complex treatments (like deep brain stimulation for Parkinson's or antiretrovirals for HIV/AIDs) and can offer themselves up as trial patients for medical researchers, thus lowering the cost and increasing the speed with which new therapies can be tested.

Many traditional trials for new therapies proceed with fewer than twenty patients in a particular experimental group (called a panel). But more than fifty thousand people use PatientsLikeMe, creating communities for particular diseases. (Its strategy is similar to PickupPal's getting drivers and riders in the same areas.) As a result, so many patients with amyotrophic lateral sclerosis (ALS, sometimes called Lou Gehrig's disease) use the site that they are actually subdivided into categories of ordinary ALS and two rarer variants: primary lateral sclerosis (PLS) and progressive muscular atrophy (PMA).

ALS affects all motor neurons in the brain and spinal cord, while PLS affects only the upper motor neurons and PMA affects only the lower ones. The distinction matters, because the average survival rate from onset is between two and five years for ALS, while it's between five and ten for PLS, and decades for PMA. The medical research community doesn't know as much about PLS and PMA, however, because as rare as ALS is (about 0.001 percent of the population suffer from it), it is about twenty times more common than PLS or PMA. The largest studies on PLS or PMA patients have examined fewer than fifty patients; by contrast, PatientsLikeMe has registered nearly two hundred with PLS and nearly three hundred with PMA.

The patients on PatientsLikeMe don't just offer themselves up for drug trials. They share their experiences with treatment (how they manage complex drug regimes) or the health-care system as a whole (how they navigate the insurance companies or Medicare). And they provide a kind of support that doctors rarely can: conversation with fellow sufferers. PatientsLikeMe uses the word *community* to denote a group of patients who share a specific condition, and for good reason: like any community of practice, they share information and ideas, and they produce cultural norms and support for one another. They offer a degree of moral support that the current medical system rarely offers, and that turns out to be a critical feature of treatment. Knowing you are not the only one going through something can itself be a huge relief, separate from any physical improvement. In addition to obtaining highly structured information about syndromes, treatments, symptoms, and so on, the patients can create their own forum topics for discussing whatever is on their minds.

Some of these conversations are involved, highly specific discussions of treatment plans. One patient reported that he got his neurologist to alter his 10mg dose of baclofen, which he was taking for "foot drop," a side effect of muscle stiffness that makes it hard to keep your balance while walking. His neurologist had told him that 10mg was the maximum dose, and he took that amount daily for fourteen years, to little effect. Then on PatientsLikeMe he read that several patients with the same condition took doses as high as 80mg without severe side effects. His doctor increased his lose, to good effect. Other conversations are sprawling, off-topic rambles—one thread, about a patient contemplating infidelity, ran to hundreds of responses (running about ten to one against). It would be easy enough to say that the discussion of baclofen is a "good" conversation on PatientsLikeMe, while the discussion of infidelity is a "bad" one, but that misunderstands not just human nature but the driving engine of the site.

PatientsLikeMe aggregates patient data better than traditional methods because it offers patients a sense of membership and shared effort. The improved quality of the data comes not in spite of conversations about infidelity and such but because those conversations and thousands of others help attact people to the site and keep them coming back. The people with ALS who inhabit PatientsLikeMe not only get things from one another that they couldn't get from professionals, they offer things the professionals couldn't otherwise get, like large populations to draw panels from.

PatientsLikeMe works because its community prizes open sharing of medical data, a cultural norm quite different from mainstream norms about medical privacy. Like most sites that handle user data, PatientsLikeMe.com has a privacy policy, but it also has an "openness philosophy":

> Currently, most health-care data are inaccessible due to privacy regulations or proprietary tactics. As a result, research is slowed, and the development of breakthrough treatments takes decades. Patients also can't get the information they need to make important treatment decisions. But it doesn't have to be that way. When you and thousands like you share your data, you open up the health-care system. You learn what's working for others. You improve your dialogue with your doctors. Best of all, you help bring better treatments to market in record time.

PatientsLikeMe offers lots of interesting tools for sharing, but the sharing itself is a human characteristic, not a technological one. As with the Invisible College's transition from alchemy to chemistry, a critical shift has taken place in the minds of PatientsLikeMe users, from a cultural norm in which medical professionals hoard information and hide it from patients to a norm of sharing, in which everyone benefits. Patients benefit from feeling and being connected, from sharing their worries and pain as well as their observations and symptoms, and researchers benefit from having the largest group of patients with chronic and rare diseases ever assembled.

PatientsLikeMe is opening up the knowledge of how to combine—it is involving patients and researchers together and making more material available for recombination. It may yet fail, but if it succeeds, it will change the culture; in fact, if it doesn't change the culture, it can't succeed, because the cultural norm that opposes sharing medical data will keep it from working.

This cultural shift is not without problems—indeed, PatientsLikeMe.com needs an openness philosophy precisely because sharing medical information entails risks, ranging from embarrassment to job discrimination to harassment. One way to get people to accept the risks of social connection is to increase its rewards; if enough people join to make the new group seem worthwhile, that will encourage still more people to join, and this feedback loop increases the value of the medical information available in aggregate. PatientsLikeMe has become so well known and appreciated that it now gets one new member for every ten new diagnoses of ALS in the United States. Not only are these patients willing to adopt the openness philosophy, but some have agreed to donate their entire genetic sequence for researchers to examine.

The story of PatientsLikeMe.com illustrates one of the most important questions we face about the uses of social media—namely, how much will we be able to take advantage of the cognitive surplus to produce real civic value?

6

Personal, Communal, Public, Civic

More value can be gotten out of voluntary participation than anyone previously imagined, thanks to improvements in our ability to connect with one another and improvements in our imagination of what is possible from such participation. We are emerging from an era of theory-induced blindness in which we thought sharing (and most nonmarket interaction) was inherently rather than accidently limited to small, tight-knit groups.

The dramatically reduced cost of public address, and the dramatically increased size of the population wired together, means that we can now turn massive aggregations of small contributions into things of lasting value. This fact, key to our current era, has been a persistent surprise. At every turn, skeptical observers have attacked the idea that pooling our cognitive surplus could work to create anything worthwhile, or suggested that if it does work, it is a kind of cheating, because sharing at a scale that competes with older institutions is somehow wrong. Steve Ballmer of Microsoft denounced the shared production of software as communism. Robert McHenry, a former editor in chief of *Encyclopedia Britannica*, likened Wikipedia to a public rest room. Andrew Keen, author of *The Cult of the Amateur*, compared bloggers to monkeys. These complaints, self-interested though they were, echoed more broadly held beliefs. Shared, unmanaged effort might be fine for picnics and bowling leagues, but serious work is done for money, by people who work in proper organizations, with managers directing their work.

Upgrading one's imagination about what is possible is always a leap of faith. In earlier eras, when amateur groups were small and organization costs were high, sharing wasn't terribly effective at creating large-scale or long-lasting value; groups were hard to coordinate, and the fruits of amateur efforts were hard to preserve, discover, or disseminate. These limits of size and longevity also limited sharing's metaphorical radius and half-life—its social radius was historically quite small, and its half-life was quite short. But social production can now be dramatically more effective than it used to be, both in absolute terms and relative to more formally managed production, because the radius and half-life of shared effort have moved from household to global scale.

This big change isn't utopia. Throwing off old constraints won't lead us to a world of no constraints. All worlds, past, present and future, have constraints; throwing off the old ones just creates a space for new ones to emerge. Increased social production heightens persistent tensions between individual and group desires. This tension was well described by Wilfred Bion, a psychotherapist who undertook group therapy with neurotics during the Second World War. Bion took notes about these sessions, which later became a slim volume called *Experiences in Groups.* During the sessions Bion observed that his patients were, as a group, conspiring to defeat therapy. They did not overtly communicate or coordinate, but whenever he tried a particular therapeutic intervention, the group would quash it by changing the subject or otherwise avoiding conversations that might lead to examining their behavior. This classically neurotic trait is usually exhibited by individuals, but Bion's group as a whole seemed neurotic as well. No one patient was doing the avoiding, or directing such responses in others, and yet some sort of coordinated response was clearly going on among the patients.

Bion wondered whether he should analyze the situation as a collection of individuals taking action, or as a coordinated group. He couldn't resolve the question, and he ultimately decided that unresolvability was the answer. To the question "Are groups of people best thought of as aggregations of individuals or as a cohesive unit?" his answer was that we are, as a species, "hopelessly committed to both." Humans are fundamentally individual, but we are also fundamentally social. Every one of us has a rational mind; we can make individual assessments and decisions. We also have an emotional mind; we can enter into deep bonds with other people that transcend our individual intellects.

All groups have an emotional component—emotion, in fact, keeps groups together. Group membership presents the individual with enough difficulties and enough opportunities to defect that without an emotional commitment, many groups would break down at the first real bit of trouble. A group that pursues a shared purpose has to be effective (or why bother to form it?), but it also has to be satisfying to its members (or why should they stay?). Groups therefore have to balance effectiveness at the group level with satisfaction at the individual level—even the Army, as hierarchically managed an institution as there is, is deeply concerned about the soldiers' morale. The question of satisfaction, though, is more significant in amateur groups, which rely more on the intrinsic motivations of their participants.

The downside of attending to the emotional life of groups is that it can swamp the ability to get anything done; a group can become more concerned with satisfying its members than with achieving its goals. Bion identified several ways that groups can slide into pure emotion—they can become "groups for pairing off," in which members are mainly interested in forming romantic couples or discussing those who form them; they can become dedicated to venerating something, continually praising the object of their affection (fan groups often have this characteristic, be they Harry Potter readers or followers of the Arsenal soccer team); or they can focus too much on real or perceived external threats. Bion trenchantly observed that because external enemies are such spurs to group solidarity, some groups will anoint paranoid leaders because such people are expert at identifying external threats, thus generating pleasurable group solidarity even when the threats aren't real.

For most groups, Bion observed, the primary threat is internal: the risk of falling into emotionally satisfying but ineffective behavior. He called groups that do so "basic groups," that is, they fall into their basest desires. Basic groups are incapable of, and often actively avoid, pursuing any higher purpose. (Bion's neurotic patients, for example, were nominally in treatment to get better but actually tried to avoid doing any work that would lead to real change.) Any group trying to create real value must police itself to ensure it isn't losing sight of its higher purpose, or what Bion called the "sophisticated goal."

By contrast, Bion called groups that pursued their goals "sophisticated work groups": their members worked to keep themselves and one another from sliding into satisfying but ineffective emotions and, when they did get sidetracked, returned the group to its sophisticated goal. The primary mechanism in such sophisticated groups is that the members internalize the standards of the group and react to behavior that undermines those standards, whether that behavior is their own or from other group members. Governance in such groups is not just a set of principles and goals, but of principles and goals that have been internalized by the participants. Such self-governance helps us behave according to our better natures.

WOMEN AND MEN

The video starts simply enough: two women, Georgia Merton and Penny Cross, sit at a table, narrating a short piece they made for Current.com about their recent trip to the beaches of France, a trip where they had stayed only with strangers.

The two coordinated their lodgings through a service called CouchSurfing.org, which is, in its own description, "a new way of traveling. With eighty thousand members, you view people's profiles and see if they can accommodate you in their homes. You don't pay, you're a guest. CouchSurfing was set up to change the way people travel. It isn't just about having free accommodations, but also about making connections worldwide." CouchSurfing is a kind of social network for travelers (now with more than one hundred thousand members) that matches people needing a place to stay for a night or two with people willing to host them.

Merton and Cross documented their use of CourchSurfing.com; their video intercut scenes of themselves traveling alongside interviews with two of their hosts, Romain in Saintes and Mounir in Biarritz. The video unfolds in chronological order, so we hear what they were thinking as they were getting ready to try couch surfing for the first time. Of course, two women traveling alone were concerned about how much trust they could put in strangers. Of the trip to Saintes, they say, "We're just looking at Romain's profile again—he doesn't look too scary in any of the photos, but you never know, do you? I mean, he could still be . . . a murderer, so we don't quite know what we're doing, and we're a bit scared about going, but we're gonna find out tomorrow." Later, before meeting Mounir and heading to his house in Biarritz, they say, "We're meeting this guy here. Georgia said earlier [in the video] that we're a bit worried, but now we're really worried. He phoned, and he only wants to take one of us at a time to his house. He says there's only one place in the car."

Romain and Mounir both turn out to be nice people—Romain, in the bit we see of him, is more introspective, talking about CouchSurfing.com and what makes it work on a social level, while Mounir is more theatrical, giving the women a mock grand tour of his apartment. But both are good hosts, and the women have a lovely time lying on the beach and hanging out with their hosts. Merton and Cross end up as converts, saying, "We're just thanking everyone we stayed with and leaving them really nice references" (a way of vouching for the men on their CouchSurfing profiles). They finish the video with a hearty recommendation to viewers to "definitely go couchsurfing!"

In real-world settings, questions of trust between men and women have always been acute. Particularly for women in an environment with men they don't know, the pleasures of novelty and social connection are balanced against both inconvenience and danger. In 2008 another pair of women, Italians Silvia Moro and Giuseppina Pasqualino di Marineo (also known Pippa Bacca), decided to make these questions of trust part of an artwork contrasting the often reflexive suspicion of others with the artists' faith in people's basic trustworthiness.

Their piece, *Brides on Tour,* had them hitchhiking around the Mediterranean, dressed only in white bridal dresses. The dresses were an emblem of purity, illustrating the commonalities of Mediterranean cultures, despite centuries of ethnic and religious tensions. Hitchhiking was also integral to the message of the piece—as the artists said on their site documenting the project, "Hitchhiking is choosing to have faith in other human beings, and man, like a small god, rewards those who have faith in him."

The pair laid out a route starting from their native Milan to Istanbul, then to Ankara, Damascus, Beirut, and Amman, ending in Jerusalem, that divided city of peace. They set out in early March, sharing photos and observations of their trip on their website as they traveled. They traveled as a pair to Istanbul; they split up to travel solo on to Beirut. Shortly after leaving Istanbul, Pippa Bacca was abducted, raped, and strangled, her body dumped behind some bushes near the town of Tavsanli. Because of the unusual and solitary nature of her travels, it wasn't immediately clear that she had gone missing, delaying the search. The last anyone heard from her was at the end of March, but her naked and decomposing body wasn't discovered until mid-April. Police arrested Murat Karatas, a local man with a criminal record, after he'd used her stolen phone. He confessed to giving her a ride in his Jeep, raping her, then killing her. The Italian and Turkish press covered the art project that had brought Pippa Bacca to Tavsanli with some delicacy, but the circumstances of her death were enough to engender a common reaction among many readers: An oddly and ostentatiously dressed woman, hitchhiking alone, in a foreign country—what was she thinking?

Two pairs of women traveled abroad, staying with strangers all the while; one pair had a pleasant vacation, the other a catastrophic one. Some of the catastrophe was bad luck—that Pippa Bacca would hitch a ride with a criminal was not foreordained—but between the artists and the couch surfers, the artists were taking the bigger risk, and they were taking it because they believed, wrongly, that human motivations are basically benign. The couch surfers, on the other hand, understood that some people have motivations to do harm, that this creates real risks, and that you have to mitigate those risks somehow. CouchSurfing.com helps travelers find hosts and vice versa, but the site also includes host profiles, a reputation system like eBay's for hosts, and lots of advice about safety, particularly for female travelers. Merton and Cross clearly thought through the risks, researched their hosts, traveled together, and took pictures and video everywhere they went. Though most of us won't take the kinds of risks that Pippa Bacca did, the general lesson is clear—increased communication and contact with others isn't risk free, and any new opportunity requires ways to manage risk. Merton and Cross managed to lower the chance of personal danger; Bacca and Moro simply denied it existed.

The couch surfers' approach reduced the danger to them, but it didn't lower danger to women overall. That much more aggressive goal requires a more coordinated approach still. In January 2009, in the southwestern Indian city of Mangalore, a group of religious fundamentalists named Sri Ram Sene attacked women drinking at Ambient, a local bar, assaulting the women and driving them out into the street. Sri Ram Sene means "Lord Ram's Army" in Hindi (Ram is a Hindu deity); this group is sometimes likened to the Taliban because of their advocacy of violent moral policing. Other pubgoers videotaped the event with their cameraphones, and those videos were in turn uploaded to YouTube and subsequently used in reports by Indian media on the attacks. The founder of Sri Ram Sene, Pramod Muthali, said they attacked the women because they were involving themselves in immoral activities: "consuming alcohol, dressing indecently, and mixing with youths of other faith." He announced that Sene's next set of victims would be anyone seen celebrating Valentine's Day, on the grounds that it was a Western celebration, inappropriate for Hindus, and that it glamorized romantic love, inappropriate for a society that prized (or in Sene's case, demanded) chaste demeanor in women.

The video of the attacks made women afraid—Sene had clearly demonstrated that it thought nothing of attacking anyone who violated social norms that Sene wanted enforced. Nisha Susan, a Mangalore resident, decided to respond by publicly rallying women to the cause. She created a Facebook group, called the Association of Pub-going, Loose and Forward Women. (Full disclosure: Susan credits my book *Here Comes Everybody* as helpful in designing her association and its response.) Susan's Facebook group acquired more than fifteen thousand members in the first few days of its existence, and their first activity was the Pink Chaddi campaign. (*Chaddi* is Hindi slang for "underwear." Sene members are called *chaddi wallahs*—underwear wearers—because members wear khaki shorts.) The idea was to publicize Pramod Muthali's mailing address and to deluge him with pink underwear, a publicly feminine gesture of the kind Sene was committed to erasing from the public arena.

Susan's campaign flooded Muthali's office with *chaddis*, many of which had confrontational messages written on them, which in turn generated more awareness of the threat Sene posed to Hindu women. The campaign had three more effects, the least important of which was the effect on Sene members themselves; they predictably vowed the pink *chaddis* wouldn't deter future actions and said they would send saris, traditional Indian dress, back to the women. (They didn't.) The second effect, more important, was that the women communicated their shared resolve to politicians in Mangalore and to the regional government of Karnataka. Unfortunately politicians and police tend to react to threats more readily if there is evidence of public concern. Participation in the Pink Chaddi campaign demonstrated publicly that a constituency of women were willing to counter Sene and wanted politicians and the police to do the same. The third and most important effect was to provide the women themselves (pub-going, loose, forward, or otherwise) with an outlet for their condemnation of both violence and repression. By demonstrating that women could create a rapid public reaction, Susan met the original Sene provocation and suggested a willingness to meet future provocations as well. In political struggles, there's no simple way to attribute cause and effect, but the state of Mangalore arrested Muthali and several key members of Sene in advance of Valentine's Day and held them until three days after as a way of preventing a repeat of the January attacks.

Both CouchSurfing.com and the Association of Pub-going, Loose and Forward Women offer ways of mitigating the specific dangers women face, but they do it in different ways. CouchSurfing is a kind of communal resource, combining individual responses into a market for surfers and surfees; its value is mainly enjoyed by its participants (and the risks are largely mitigated by its participants as well). Susan's association, by contrast, was a civic intervention, designed to make India safer not just for the women who mailed the *chaddis* but for all women who want to be free of the threat posed by Sri Ram Sene. The differing methods and results of these two groups illustrate ways that voluntary participation can change society.

INDIVIDUALS, GROUPS, AND FREEDOM

Today people have new freedom to act in concert and in public. In terms of personal satisfaction, this good is fairly uncomplicated—

even the banal uses of our creative capacity (posting YouTube videos of kittens on treadmills or writing bloviating blog posts) are still more creative and generous than watching TV. We don't really care how individuals create and share; it's enough that they exercise this kind of freedom.

Increases in personal satisfaction, though, are not all that's at stake. In terms of social, as opposed to individual, value, we care a lot about how our cognitive surplus gets used. Participating in Ushahidi creates more value for society than participating in ICanHasCheezburger; making and sharing open source software creates value for more people than making and sharing Harry Potter fan fiction. The value from Ushahidi or open source software is more than the sum of the personal satisfactions of the participants; nonparticipants also derive value from the effort. You can think of this scale of value as rising from personal to communal to public to civic.

Personal value is the kind of value we receive from being active instead of passive, creative instead of consumptive. If you take a photo, or weave a basket, or build a model train set, you get something out of the experience. This energy drives the world's hobbyists. However, as Katherine Stone (the medical advocate quoted in Chapter 3) notes, there's great value in seeing that we are not alone. Adding the social motivations of membership and generosity to the personal motivations of autonomy and competence can dramatically increase activities. Now that people can share videos on YouTube, far more people make such videos than ever made them when sharing them was harder and the potential audience smaller. Because humans have fundamentally social as well personal motivations ("Hopelessly committed to both," as

Bion said), social motivations can drive far more participation than can personal motivations alone.

The spread of social media that allows for public address has led to a subtle change in the word *share.* Sharing has typically required a high degree of connection between donor and recipient, so the idea of sharing a photograph implied that you knew the sharees. Such sharing tended to be a reciprocal and coordinated action among people who knew one another. But now that social media has dramatically lengthened the radius and half-life of sharing, the organization of sharing has many forms.

While these various forms exist on a spectrum, we can identify four essential points on that spectrum. One such form is personal sharing, done among otherwise uncoordinated individuals; think ICanHasCheezburger. Another, more involved form is communal sharing, which takes place inside a group of collaborators; think Meetup.com groups for post-partum depression. Then there is public sharing, when a group of collaborators actively wants to create a public resource; think the Apache software project. Finally, civic sharing is when a group is actively trying to transform society; think Pink Chaddi. The spectrum from personal to communal to public to civic describes the degree of value created for participants versus nonparticipants. With personal sharing, most or all of the value goes to the participants, while at the other end of the spectrum, attempts at civic sharing are specifically designed to generate real change in the society the participants are embedded in.

Personal sharing is the simplest kind; both the participants and the beneficiaries are acting individually but get personal value out of one another's presence. Digital tools create a long-term potential for sharing with no additional requirements for the sharer or the sharee. Sharing a photo by making it available online constitutes sharing even if no one ever looks at it.

This "frozen sharing" creates great potential value. Enormous databases of images, text, videos, and so on include many items that have never been looked at or read, but it costs little to keep those things available, and they may be useful to one person, years in the future. That tiny bit of value may seem too small to care about, but with two billion potential providers, and two billion potential users, tiny value times that scale is huge in aggregate. Much creative energy that was previously personal has acquired a shared component, even if only in frozen sharing.

Creating communal value is more complicated. A buildup of uncoordinated contributions can create personal value, but a group of people conversing or collaborating with one another can create communal value. The Meetup.com groups for postpartum depression create value for their members. Most of that value, though, is enjoyed by the Meetup members themselves. Communal value requires more interaction than personal value, but that value still remains within the circle of participants.

Public value is as interactive as communal value, but far more open to participation from and sharing with newcomers and outsiders. In contrast to communal sharing, public sharing allows people to join in at will, and the results will be made available even to those members of the public who are not participants. The Apache project is a classic example of sharing that creates public value, because the programmers creating the Apache project generally use it themselves but also want (and benefit from) adoption by the outside world. Indeed, a key motivation for many participants in open source projects is a positive view of the kind of public value such projects can create.

Civic value is like public value in its openness, but for groups dedicated to creating civic value, improving society is their explicit goal. Nisha Susan's Association of Pub-going, Loose and Forward Women was designed to improve the freedom of all Indian women, not just the members. If the association had created value only for its members, in the manner of communal sharing, it would have been a failure. Civic participants don't aim to make life better merely for members of the group. They want to improve even the lives of people who never participate (with the obvious caveat that improving the freedom of women makes life worse for people like Pramod Muthali, who object in principle to that kind of freedom).

These different kinds of participation don't mean that we should never have lolcats and fan fiction communities—it's just that anything at the personal and communal end of the spectrum isn't in much danger of going away, or even of being underprovisioned. It's hard to imagine a future where someone asks himself, "Where, oh where can I share a picture of my cute kitten?" Almost by definition, if people want that kind of value, it will be there. It's not so simple with public and especially civic value. As Gary Kamiya has noted of today's web, "You can always get what you want, but you can't always get what you need." The kinds of things we need are produced by groups pursuing public value.

We should care more about public and civic value than about personal or communal value because society benefits more from them, but also because public and civic value are harder to create.

The amount of public and civic value we get out of our cognitive surplus is an open question, and one strongly affected by the culture of the groups doing the sharing, and by the culture of the larger society that those groups are embedded in. As Dean Kamen, the inventor and entrepreneur, puts it, "In a free culture, you get what you celebrate." Depending on what we celebrate in one another, we can get a few pieces of public and civic value, like those we see today in Wikipedia and open source software and the Responsible Citizens, or we can celebrate people who create civic value, making it a deep part of the experience of users everywhere.

Getting what we celebrate highlights the tension between maximizing individual freedom and maximizing social value. Social media introduces social dilemmas into a number of environments where they didn't previously exist; prior to the present historical generation, motivating unpaid actors to do anything for the civic good was left to governments and nonprofits, themselves institutional actors. Today we can take on some of those problems ourselves, but the more we want to do so at the civic end of the scale, the more we have to bind ourselves to one another to achieve (and celebrate) shared goals. Neither perfect individual freedom nor perfect social control is optimal (Ayn Rand and Vladimir Lenin both overshot the mark), so it falls to us to manage the tension between individual freedom and social value, a trade-off that follows the by-now-familiar pattern of having no solution, just different optimizations that create different kinds of value, and different kinds of problems that need to be managed.

GROUPS AND GOVERNANCE

Sharing thoughts and expressions and even actions with others, possibly many others, is becoming a normal opportunity, not just for professionals and experts but for anyone who wants it. This opportunity can work on scales and over durations that were previously unimaginable. Unlike personal or communal value, public value requires not just new opportunities for old motivations; it requires governance, which is to say ways of discouraging or preventing people from wrecking either the process or the product of the group.

Pierre Omidyar, the founder of eBay, has often said that he built that auction site on the assumption that "people are basically good," meaning that most transactions on the site would go well if he let buyers and sellers find each other. This idea, noble as it is, didn't quite pan out for eBay—within weeks of its launch, so many transactions were marred by cheating that eBay began a reputation system wherein buyers and sellers could get a sense of one another's honesty, promptness, and so on, based on reviews by other members. Ultimately members' reputations mattered enough to keep fraud to a minimum. Buyers and sellers with long-term identities and reputations on the site were provided with an incentive not just to behave well but to be seen as behaving well. Paul Resnick, a social media researcher at the University of Michigan, studied eBay's reputation system and concluded that sellers with a positive reputation, as reported by their customers, could command an 8 percent premium on price over sellers who'd just arrived on the site.

Omidyar's original dictum—"People are basically good"—is

true only with some commitment to governance structure. EBay, CouchSurfing, PickupPal, and countless other sites that involve real effort or money, as well as real risk, have had to find ways to govern their members in order to produce a larger good. The less catchy but more accurate lesson from eBay is "People will behave if they sense that there is long-term value in doing so, and short-term loss in not doing so." The greater the value and the risk inherent in participation, the more some sort of structure is required to keep the participants concentrating on their shared and sophisticated goals, rather than on their personal and basic ones.

There is no one-size-fits-all set of rules for governing groups that create public value. Working software projects like Apache tend to be brutal technical meritocracies, while groups coordinated via social networks, like the Responsible Citizens, tend to have a more supportive culture, and so on. There are two universals, however: to succeed in creating and sustaining public value, a group must defend itself against external threats (like eBay defending itself against fraud) and against internal threats (like the members of the Apache project defending themselves against getting sidetracked by arguments or inertia). As Bion noted, external threats are the more attention-getting, as groups can easily focus their energies on external enemies, but when it comes to keeping a group of voluntary participants committed to the creation of shared value, internal threats are far more serious.

It's easy to galvanize a group with thoughts of some external enemy, but, in fact, the likeliest source of distraction from a shared goal is from the members of the group with that goal. (Ironically, one of the easiest ways to distract such a group is to get them to focus on outside enemies, real or imagined, rather than on their shared interests or tasks.) Because the biggest threat to group action is internal, voluntary groups need governance so that we can defend ourselves from ourselves; we need governance to create a space we can create in. Creativity at the personal and communal end of the spectrum requires little of that sort of governance to survive, but the more a group wants to take on hard public or civic problems, the greater the internal threats of distraction or dissipation are and the stronger the norms of governance need to be.

Falling costs create room for experimentation, experiments create value, and that value creates an incentive to benefit from it. If incentive led only to more experimentation, then lowered costs would create a pure virtuous circle. Unfortunately, the incentive to make use of experimental value reaches people who had nothing to do with creating or sustaining it. The larger and more publicly successful a project is, the more people will want to appropriate that value while giving nothing back or even to see the project fail.

To take a participatory example that suffers from these vicious cycles in several ways, consider Wikipedia. Some people act out on Wikipedia to get attention. Shane Fitzgerald, a twenty-two-year-old Dublin student, added a fake quote to composer Maurice Jarre's Wikipedia page, from whence it turned up in Jarre's obituaries worldwide. Other times people want to alter or silence a point of view they dislike. Unlike Fitzgerald's playful hoax, Wikipedia pages on subjects ranging from evolution to Islam to Microsoft to Galileo are under fairly steady threat from people who want the contents significantly altered or removed. Sometimes the appropriation is an attempt to capture financial value, as when people try to edit Wikipedia articles to add favorable statements about a particular company or to remove unfavorable ones. This tension between the individual and the group reflects the strains involved in taking advantage of cognitive surplus for public and civic uses.

The choice we face is this: out of the mass of our shared cognitive surplus, we can create an Invisible University—many Invisible Colleges doing the hard work of creating many kinds of public and civic value—or we can settle for Invisible High School, where we get lolcats but no open source software, fan fiction but no improvement in medical research. The Invisible High School is already widespread, and our ability to participate in ways that reward personal or communal value is in no imminent danger. Following Gary Kamiya's observation about the ease of getting what we want, we can always use the internet today to find something entertaining to read, watch, or listen to.

Creating real public or civic value, though, requires more than posting funny pictures. Public and civic value require commitment and hard work among the core group of participants. It also requires that these groups be self-governing and submit to constraints that help them ignore distracting and entertaining material and stay focused instead on some sophisticated task. Getting an Invisible University means mastering the art of creating groups that commit themselves to working together outside existing market and managerial structures, in order to create opportunities for planet-scale sharing. This work is not easy, and it never goes smoothly. Because we are hopelessly committed to both individual satisfaction and group effectiveness, groups committed to public or civic value are rarely permanent. Instead, groups need to acquire a culture that rewards their members for doing that hard work. It takes this kind of group effort to get what we need, not just what we want; understanding how to create and maintain it is one of the great challenges of our era.

7

Looking for the Mouse

This book is about a novel resource, and about ways society might take advantage of it. That resource isn't just our cumulative free time, which first ballooned with the rise of the forty-hour workweek and grew after the Second World War with larger, healthier populations, rising educational opportunities, and spreading prosperity. All that free time wasn't yet a cognitive surplus, because we lacked the means to make use of it. In fact, even as the developed world's cumulative free time was growing, many of the older social structures that brought us together were dismantled, ranging from picnics and neighborhood associations to bowling leagues and pedestrian shopping. That shrinking of participatory options made managing our free time a largely personal issue, more a matter of using it up than actually using it.

Nor is the current transformation of free time into cognitive surplus just about new social tools. Although media that supports public participation, sharing, and discussion is a novelty, merely having the means to share without the motive to do so doesn't mean much. Any voluntary activity has to offer opportunities that tap some real human motivation. If I told you I'd created a tool to help the alumni association of your college find you more easily, you might not think much of my invention. If, on the other hand, I told you I'd invented a tool for having the alumni themselves (aka your old college friends) get in touch with you, you might consider the proposition differently. As we know from the rise of social networking services like Facebook, Twenty in Spain, or QQ in China, this kind of connection is one of the world's most popular uses of social media.

The fusing of means, motive, and opportunity creates our cognitive surplus out of the raw material of accumulated free time. The real change comes from our awareness that this surplus creates unprecedented opportunities, or rather that it creates an unprecedented opportunity for us to create those opportunities for each other. The low cost of experimentation and the huge base of potential users mean that someone with an idea that would require dozens (or thousands) of participants can now try it, at remarkably low cost, without needing to ask anyone for permission first.

All this has happened already. It's novel and surprising, but the basic change is complete. What has not yet happened, what is in fact still an open question, is what benefit will eventually emerge from our ability to treat the world's cognitive surplus as a shared and cumulative resource. Given the explosion of creative and generous behaviors, we might assume that good uses of that surplus will just happen. This is true, but only for some of those potential uses.

The world is becoming well provisioned with sources of personal and communal value, value mainly created and captured by the participants. Down at the lolcats end of the spectrum, the current experimentation is unlikely to stop anytime soon. At the civic end of the spectrum, though, we can't count on new kinds of socially beneficial activities just happening. Creating a participatory culture with wider benefits for society is harder than sharing amusing photos. How much of that social change are we going to be able to grasp?

The earliest visibly successful uses of cognitive surplus were in technical communities of computer programmers, where collaborative behaviors are well understood and where cultural barriers to participating are few. Programmers who work on open source projects like Apache and Linux are by definition people who view participation in a positive light. Steve Weber, a Berkeley political scientist and one of the great chroniclers of the open source movement, notes in his book *The Success of Open Source* that neither the reduced costs of collaboration nor the eventual technical quality of the output can fully explain a person's choice to work on an open source project. Instead, a critical mass of core programmers have to have "a positive normative or ethical valence toward the process," which is to say, to have made a deep judgment that social production is the right way to create software. (This is a practical version of Dominique Foray's observations from Chapter 5, that the value of combinability, what programmers do daily, is strongly affected by culture.)

The open source model for shared creation has spread to many nontechnical domains, from carpooling to patient support groups, but civic-mindedness doesn't automatically flow from contemporary culture. A cynical streak in society looks at all forms of amateur participation as either naive or stupid. (My teenage self tapped into that strain of contempt when I was thinking about other people's hobbies.) It's tempting to imagine a broad conversation about what we as a society should do with the possibilities and virtues of participation.

Such a conversation will never happen. If you do a web search for "we as a society," you will find a litany of failed causes, because society isn't the kind of unit that can have conversations, come to decisions, and take action. Civic value rarely comes from sudden social conversions; nor does it bubble up from individual actions. It comes, instead, from the work of groups, small groups at first that grow in size and importance, the pattern of collaborative circles, communities and practice, and many other group patterns. If we want to create new forms of civic value, we need to improve the ability of small groups to try radical things, to help the inventors of the next PatientsLikeMe or the next set of Responsible Citizens get up and going. It's from groups trying new things that the most profound uses of social media have hitherto come and will come in the future.

The essential source of value right now is coming less from master strategy than from broad experimentation, because no one has a complete grasp, or even a very good one, about what the next great idea will look like. We are all living through the disorientation that comes from including two billion new participants in a media landscape previously operated by a small group of professionals. When this much has changed, our best chance for finding good ideas is to have as many groups as possible try as many things as possible. The future doesn't unfold on some preordained track; things change because someone figures out something that is possible right now, and pushes to make it happen.

PARADOX OF REVOLUTION

Johannes Gutenberg's best-known work was his forty-two-line Bible, a spectacularly beautiful example of early printing. But it was neither his first work nor his most voluminous. (He printed fewer than two hundred copies.) That honor goes instead to his printing of indulgences.

An indulgence, in Catholic theology, is a way to reduce the amount of time a person spends in purgatory for sins that have already been forgiven. Sinning, Catholics believe, runs up the time you have to wait after death to get into heaven. Indulgences are a way to reduce that wait, and the way you get an indulgence is to make a donation to the Church. The practice was viewed with suspicion in some theological quarters as an exchange of value that veered dangerously close to a purchase, but so long as the exchange of indulgences for donations was allowed, the desire to both issue and receive them was also there.

In Gutenberg's time, an indulgence was a written record of the transaction, which also worked as a kind of token that validated the bearer's future time off. The Church would deputize people to issue indulgences and collect money on its behalf; the issuer got a cut of the proceeds for his trouble. (Chaucer's "The Pardoner's Tale" is told by one of these issuers.) However, the income from indulgences was constrained by the speed at which they could be written out by hand. The result was an imbalance of supply and demand; the world wanted more indulgence than the Church could provide.

Enter Gutenberg. He got a loan to enter the indulgence-printing business, and by typesetting them, he was able to dramatically increase the supply, expanding both the market and his cut of it. He printed indulgences, probably by the thousands (few have survived) before he printed his first Bible. (One source suggests he originally had to print his Bibles in secret, as he'd secured his loan for the much more profitable work of indulgences.) If you had seen Gutenberg's shop in the 1450s, when its output was indulgences and Bibles, you might have thought the printing press was custom-made for strengthening the economic and political position of the Church. And then a funny thing happened: just the opposite.

Gutenberg's press flooded the market. In the early 1500s John Tetzel, the head pardoner for German territories, would sweep into a town with a collection of already printed indulgences, hawking them with a phrase usually translated as "When a coin a coffer rings / A soul for heaven springs." The nakedly commercial aspects of indulgences, among other things, enraged Martin Luther, who in 1517 launched an attack on the Church in the form of his famous *Ninety-five Theses*. He first nailed the theses to a church door in Wittenberg, but copies were soon printed up and disseminated widely. Luther's critique, along with the spread of Bibles translated into local languages, drove the Protestant Reformation, plunging the Church (and Europe) into crisis.

The tool that looked like it would strengthen the social structure of the age instead upended it. From the vantage point of 1450, the new technology seemed to do nothing more than offer the existing society a faster and cheaper way to do what it was already doing. By 1550 it had become apparent that the volume of indulgences had debauched their value, creating "indulgence inflation"—further evidence that abundance can be harder for a society to deal with than scarcity. Similarly, the spread of Bibles wasn't a case of more of the same, but rather of more is different—the number of Bibles produced increased the range of Bibles produced, with cheap Bibles translated into local languages undermining the interpretative monopoly of the clergy, since churchgoers could now hear what the Bible said in their own language, and literate citizens could read it for themselves, with no priest anywhere near. By the middle of the century, Luther's Protestant Reformation had taken hold, and the Church's role as the pan-European economic, cultural, intellectual, and religious force was ending.

This is the paradox of revolution. The bigger the opportunity offered by new tools, the less completely anyone can extrapolate the future from the previous shape of society. So it is today. The communications tools we now have, which a mere decade ago seemed to offer an improvement to the twentieth-century media landscape, are now seen to be rapidly eroding it instead. A society where everyone has some kind of access to the public sphere is a different kind of society than one where citizens approach media as mere consumers.

The early print revolution also reminds us that at the beginning of the spread of a new tool, it is too early to know how (and where and how much) society will change because of its use. Big changes can stall. After the initial spread of indulgences, the increased volume of their production dramatically decreased their value. Small changes can spread. The *Ninety-five Theses*, nailed to a single door, were reprinted and translated and reprinted again, spreading far and wide. What seems to threaten uniformity actually creates diversity. As Elizabeth Eisenstein notes in *The Printing Press as an Agent of Change*, observers of early print culture assumed that the abundance of books would mean more people reading the same few texts. The press seemed to offer (or threaten, depending on your point of view) an increase in monoculture as a small group of books would become the shared literary patrimony of a whole continent. As it turned out, the press undermined rather than strengthened the earlier intellectual culture. Because each reader had access to more books, intellectual diversity, not uniformity, was the result. This increase in diversity of sources corroded faith in older institutions. When a scholar could read both Aristotle and Galen side by side and see that the two sources clashed, it corroded reflexive faith in the ancients. If you couldn't trust Aristotle, who could you trust?

The changes today have something of that feeling. When the general public began using digital networks, the idea that everyone would contribute something to the public sphere was assumed to be contradictory to human nature (for which read: accidental behaviors of the twentieth century). And yet our desire to communicate with one another has turned out to be one of the most stable features of the current environment. The use of tools that support public expression has gone from narrow to broad in the space of a decade. What seemed a new channel for traditional media is actually changing it; what seemed to threaten cultural uniformity is actually creating diversity.

Most of the world's adults now use digital networks, whether via computer or phone, and most of those began doing so only in the last decade. Observers of society had a fairly unprecedented opportunity to observe people's behavior around the adoption of digital tools, and the result is exactly what you'd expect from the arrival of an unfamiliar new medium: we are absolutely terrible at predicting our own future behavior. Study after study in the 1990s asked potential users what they would do with the internet if they got access to it, and the commonest answers always clustered around "I'll use it to find information," "I'll use it to help me with my schoolwork," and so on. Whenever a poll asked people already online what they actually did, the answers were quite different. "Keeping up with friends and family," "sharing photos with others," "talking with people who share my interests," and the like appeared near the top of every list. Because we're so lousy at predicting what we will do with new communications tools before we try them, this particular revolution, like the print revolution, is being driven by overlapping experiments whose ramifications are never clear at first. Hence creating the most value from a tool involves not master plans or great leaps forward but constant trial and error. The key question for any society undergoing such a shift is how to get the most out of that process.

The possibility of large-scale sharing—massive, continual sharing among various groups drawn from a potential pool of two billion people, is already manifesting itself in many places, from the globalization of charity to the logic of higher education to the conduct of medical research. The opportunity we collectively share, though, is much larger than even a book's worth of examples can express, because those examples, and especially the ones that involve significant cultural disruption, could turn out to be special cases. As with previous revolutions driven by technology—whether it is the rise of literate and scientific culture with the spread of the printing press or the economic and social globalization that followed the invention of the telegraph—what matters now is not the new capabilities we have, but how we turn those capabilities, both technical and social, into opportunities. The question we now face, all of us who have access to new models of sharing, is what we'll do with those opportunities. The question will be answered much more decisively by the opportunities we provide for one another and by the culture of the groups we form than by any particular technology.

IMPROVING THE ODDS

The earliest medium that provided a platform for distributed group conversation was a computer system called PLATO that launched in Minnesota in the early 1960s. The system was slow and text-only, but it still offered real human interaction. People quickly used this experiment in electronic education for the whole range of social experiences possible in online spaces: friends (and enemies) were made, relationships started (some leading to marriage), and arguments among the users sprang up, died down, and sprang up again. PLATO was the first place where all these effects could be seen in a digital environment.

Much about social media has changed since then—computers are faster, phones are more capable, networks are better—but from PLATO to today, there are two unbroken lines of thought about the social uses of communications tools. The first is that users never behave exactly as the designers of the system expect or want them to. That was as true of PLATO as it is of Facebook. The second is that observers have a desire to bring the complexity to heel by creating a recipe for creating a successful community.

Sadly, such a recipe isn't possible. Social systems are complex, not just because of software features or even social interactions, but because of cultural context. The first social networking service wasn't Facebook in 2004 or Friendster in 2002 but a service called SixDegrees.com, launched in 1996. Six Degrees failed to become a viable social network, not because its technology was wrong while Friendster's was right, but because in 1996 not enough people were comfortable living their social lives online. Similarly, YouTube was only one of many video-sharing services in 2005, when it was used to share the popular music video "Lazy Sunday." Whatever YouTube's technical advantages over its competitors, it became synonymous with video sharing in part because of its lucky break in being the service used to host that video. That input to its success was user-driven and accidental rather than technological and planned.

With social software, there are no foolproof recipes for success. (I speak from bitter experience, having participated in the creation of both successful and failed social media.) And yet we've learned some things about human interaction in the last few decades. The trick for creating new social media is to use those lessons to weight the odds in your favor, rather than as a set of instructions that guarantees success. With that caveat, I offer some of those lessons, as ways to improve odds for successful harnessing of cognitive surplus. I've broken the observations into three categories: creating new opportunities, dealing with early growth, and adapting as the users generate surprises along the way.

Starting

You can never get complex social interactions right first crack out of the box, but you can get them wrong. The key to starting well is to understand how the initial launch of social media is special.

—Start Small

It's easy to imagine a service that will be useful if a lot of people are using it; it's hard to create a service that will be useful when only a few people are using it. Imagine PickupPal.com with only one hundred users spread out over the same area that the service covers today—one driver in Ottawa and one rider in Oslo and so on. It would have been a disaster. Its creators' solution was to recruit drivers and riders in Ontario first, to show that the service could work, then move outward from there. Similarly, the Apache webserver did not start with thousands of programmers working on it; it started with half a dozen, and only when those people made something worth looking at did it double in size, and only when those people made something worth looking at did it double again.

Projects that will work only if they grow large generally won't grow large; people who are fixated on creating large-scale future success can actually reduce the possibility of creating the small-scale here-and-now successes needed to get there. A veritable natural law in social media is that to get to a system that is large and good, it is far better to start with a system that is small and good and work on making it bigger than to start with a system that is large and mediocre and working on making it better.

—Ask "Why?"

Individual people have different motivations for doing things, and those motivations create different logics of participation. Some work well together (competence and membership are both rewarded by being part of a collaborative circle; autonomy and generosity are both rewarded by writing open source software). Some can be at cross-purposes (autonomy can be in tension with membership, whenever doing something by yourself feels different from doing it with others). Some can even crowd out others (paying users to sell things to one another, as with Amway or Avon, can crowd out intrinsic motivations to participate).

Even knowing what intrinsic motivations are, we can't predict how people will react to a given opportunity. Why would users care about this particular opportunity, given all the other things they could be doing with their time? New ideas seem clearer and more obviously good to the founders and designers of a service than to potential users, and designers can easily imagine users happily acting in a way that matches their goals. (Remember Hank, the Angry Drunken Dwarf?) Designers of these services have to put themselves in the user's position and take a skeptical look at what the user gets out of participating, especially when the motivation of the designer differs from that of the user.

—Behavior Follows Opportunity

Behavior is motivation that has been filtered through opportunity. Even after you decide why users will want to participate in your new service, you have to give them an opportunity to do so in a way they can understand and care about. This is hard, because you can't just present them with generic capability. Every social media user can already create any number of things online, whether it's a piece of writing or a photo or a video, and they can join any number of online communities dedicated to discussion of things they care about. Given this riot of opportunity, you have to give your users a specific one that rewards their intrinsic motivations, preferably both personal ones (like autonomy and competence) and social ones (like membership and generosity).

As Joshua Porter, a social media designer who writes the influential weblog Bokardo, explains to his clients, "The behavior you're seeing is the behavior you've designed for." Users will only take advantage of opportunities they understand and that seem interesting or valuable. Porter is in effect telling his clients: It doesn't matter how much you want users to behave a certain way. What matters is how they react to the opportunities you give them. If you want different behavior, you have to provide different opportunities.

—Default to Social

Back in 2003, a service called Delicious.com offered users a way to save webpages they'd found, adding labels and notes so they could organize those pages. Delicious created value for its users in two ways. First, it let each user find and remember particular pages on the web, and second, it let all users look through others' collections of remembered webpages as well. The service grew quickly, rising from dozens of users to millions in the space of a couple of years. Storing personal lists of webpages wasn't a new idea, however. Back in the 1990s a company called Backflip.com offered the same service; unlike Delicious, however, Backflip failed to attract a significant number of users.

So what was the difference? Backflip.com assumed that the personal utility was paramount; it provided an option for users to share their bookmarks, but users had to opt into it, which few did. Delicious, by contrast, didn't provide this option; it always shared all your bookmarks. (It later added private bookmarks, but only after it achieved success as a "public-only" service.) As Kevin Kelly noted in his piece "Triumph of the Default" (see Chapter 4), the careful use of defaults can shape how users behave, because they communicate some expectation (the expectation has to be one the users are happy to follow). Backflip concentrated on personal value and assumed social value was optional. Delicious, on the other hand, made social value the default. By assuming that users would be happy to create something of value for each other, Delicious grew quickly, since the social value attracted new users, and their subsequent use of the service created still more social value.

Growing

Social systems have two modes—dynamic and dead. Even stable social systems are only relatively stable, as the users are constantly interacting with one another, and with the system. One of the great challenges of such systems, especially in their early days, is to manage the dynamics of growth.

—A Hundred Users Are Harder Than a Dozen and Harder Than a Thousand

It's easy to see how a social service with only a dozen users could work well. The users could all have their say, could know something of one another's personalities and quirks, and could rely on the group's small size to keep the worst public argumentation at bay. It's also easy to imagine a social service with a thousand users. Having that many users would give a service all kinds of participants: highly active and completely passive participants, boosters and critics, arguers and peacemakers, and so on. Amid all their individual and seemingly chaotic interactions, however, users of such systems actually exhibit a surprising continuity of commitment.

However, it's surprisingly hard to operate between the intimate scale of dozens and the public scale of thousands or more. This middle scale, something like a hundred people, is often too big to operate as a single group but too small to become socially self-sustaining. It roughly marks the transition from relatively balanced to dramatically unbalanced participation. Being a participant in a midsize group often feels lousy, because you get neither the pleasures of tight interconnection nor the advantages of urban scale and diversity.

The figure one hundred is a guide rather than a rule; different services transition at different sizes, but trade-offs of scale are built into social systems generally. The key transition is around culture. A small group where everyone knows everyone else can rely on personality to arrange its affairs, while a large group will have some kind of preexisting culture that new users adopt. In the transition between those two scales is where culture gets established. (By aligning members' actions and assumptions, even when they don't know each other, culture is a way to keep the growing complexity of large groups at bay.) Once a culture is established, whether it's helpful or suspicious, accepting or skeptical, it is very hard to change.

The key is to recruit as the first dozens of users people who will embody the right cultural norms, with the caveat that what makes a set of norms right differs from place to place. A technical project like Apache will need early users with technical talent and a willingness to argue; a social project like the Responsible Citizens will need positive deviants; and so on. No one kind of user, and no one kind of culture, is right for all environments, but whatever culture has taken hold by the time you get to a hundred users has a good chance of remaining in force when you get to a thousand (or a million).

—People Differ. More People Differ More.

The twentieth century inculcated in us "the myth of the audience," the notion that people are generally the same, and that any large group of readers, listeners, or viewers is a relatively uniform lump of consumers. (In this view, knowing whether someone is a teenage boy or a middle-aged woman constitutes a highly refined distinction.) Where the myth of the audience held true, it was true largely by accident.

People's behaviors as consumers, when given a narrow range of choices, do indeed converge. When media channels are limited and cost of production is high, representation of interests is limited. But when anyone can create media, and media helps coordinate the former audience, the array of interests is dizzying. (We might call this the Mirzoeff principle, after my NYU colleague quoted in Chapter 3, who observed the "full crazy range" of interests we can now find on the web if we go looking for them.)

In broadcast systems, the bigger the group, the more behavior converges on some kind of average; in participatory systems, "average" is an almost useless concept. The behaviors of the most active and least active members diverge sharply as the population grows. The bigger the social system, the more dramatic the difference between the most active and least active participants. In small groups everyone can participate roughly equally, but in large systems a core group and a peripheral one emerge (the pattern of collaborative circles). The larger the system, the larger the difference in involvement between the core and the peripheral members. Large populations have a greater range of behavior than small ones, making the idea of "an average user" less and less useful as the system grows.

People building or running a social service can't insist that participation be either equal or universal; it won't be either unless the participant population is kept small. Instead, the service can take advantage of this divergence by offering different levels of involvement to different users. Wikipedia offers potential participants the ability to do as much writing or editing as they like, but also as little. If you fix a typo and never do anything on Wikipedia again, that still has more value than if you hadn't fixed it. Wikipedia makes it as easy as possible to effect these small changes, not even making users set up an account before they start editing. This low threshold to participation invites the accumulation of the smallest units of value—no one would create an account just to fix a single typo. By making the size of the smallest possible contribution very small, and by making the threshold for making that change small as well, Wikipedia maximizes contributions across an enormous range of participation. This wouldn't have worked when amateur participation was limited, but it works remarkably well when the participant pool can be drawn from the whole world.

—Intimacy Doesn't Scale

You can have an intimate dinner party for six but not for sixty. More is different, and in social settings that difference expresses itself in the logic of clusters. In a small group, everyone might be tightly connected to everyone else. But as the system grows, that possibility disappears; either the participants become an audience or they cluster into small, overlapping groups that preserve some intimacy.

In an audience, everyone sees the same thing; at large scale, even sites that seem to offer the possibility of interaction are actually little more than broadcast outlets, with a thin icing of participation. CNN.com's "Sound Off" feature allows readers to comment on its articles. The site has millions of readers, but most of the articles generate only a few dozen comments, and a rare few generate hundreds. Better than 99 percent of the audience members don't participate, they just consume, and most of those who do leave comments are sounding off rather than conversing. This model differs from the anonymously consumed public media of the twentieth century, but not much.

At the other extreme, some services with millions of participants allow those participants to cluster into many smaller, socially denser groups. Yahoo.com hosts millions of mailing lists, to which tens of millions of people subscribe, but people are either on a mailing list or they are not—the lines around the individual clusters are clearly drawn. Few of those millions of users think of themselves as being part of a larger Yahoo community, even though Yahoo is their host. Their allegiance is to the local cluster of people on their mailing list.

Facebook is in the middle of this audience-and-cluster spectrum. Facebook doesn't have a single center, as CNN.com does, nor a set of sharply drawn edges, as mailing lists do. Instead, it has overlapping social horizons. Facebook says it has more than three hundred million users, but none of them experiences being part of a group of three hundred million. Instead, Facebook users cluster into much smaller groups, with dozens of friends. Those clusters are considerably more involved with one another than any random sampling of the CNN audience (or of CNN commentators), but they are considerably less involved than members of a small mailing list.

Every service that wants to harness the cognitive surplus at large scale faces these trade-offs. You can have a large group of users. You can have an active group of users. You can have a group of users all paying attention to the same thing. Pick two, because you can't have all three at the same time.

—Support a Supportive Culture

Amtrak, the U.S. passenger rail company, has on many of its trains a "quiet car." The rules are fairly self-explanatory: no music without headphones, no loud talking, and no mobile phone conversations. It's that last one that trips people up.

Businessmen (they are always men in my experience) violate this rule with some regularity, either not knowing that they are on the quiet car or forgetting when they reflexively reach for their phone. Remarkably, the other passengers react quickly and publicly, shushing offending phone users within seconds. They even take a certain pleasure in policing the rules of the quiet car, the same effect observed in the Ultimatum Game, when responders are willing to expend resources to punish ungenerous proposers. Given the many examples of rudeness and mobile phone use in public, what's special about the quiet car? It's that the riders know they can call for backup.

The quiet in the quiet car is one of Elinor Ostrom's collective resource problems. The riders are willing to police the rules themselves, because they know that if an argument ensues, the conductor will appear and take over enforcement. A visible willingness to enforce the rules, in other words, actually reduces the amount of energy the people who run the train have to expend on policing, because the riders are willing to coordinate a response among themselves, knowing they can count on predictable support. (One of the most parsimonious examples of this pattern on the web is from JavaRanch, a site for people learning the Java programming language; one of the rules for participants on the site reads, in full, "Be nice.")

Adapting

No one gets it right the first time. Wikipedia was born out of the ashes of a previous, failed experiment in creating an online encyclopedia called Nupedia. Twitter was created for use on mobile phones, then retooled itself for more web use, then saw use explode with the spread of smartphones. If successful uses of cognitive surplus required designers to get it right the first time, you'd be able to count the successes on the fingers of one hand. Instead, the imperative is to learn from failure, adapt, and learn again.

—The Faster You Learn, the Sooner You'll Be Able to Adapt

The possibilities for continual learning with social media are dramatic. When the photo-sharing service Flickr.com was experimenting most actively with new features, it sometimes upgraded its software every half hour, at a time when traditional software upgrades were released annually. Meetup.com, the service that helps people gather in like-minded groups in their local communities, has its designers watch people trying to use their service every day, instead of having a focus group every six months.

Organizations in the twentieth century used all kinds of proxy measures to study what their customers or patrons or users were doing, things like focus groups and surveys. Those methods helped understand user motivations directly, but many of the difficulties in understanding have vanished. Organizations most often fail to learn from their users because of a bias toward the "office drone/couch potato" view of humanity, but successful uses of cognitive surplus figure out how to change the opportunities on offer, rather than worrying about how to change the users.

—Success Causes More Problems Than Failure

Outright failure, always a possibility with new social media, is at least a clean case. The true long-term difficulties come from success, because successful services heighten expectations and attract people who want to take advantage of the goodwill of others (by doing things like sending spam) or see the project fail (as with the bus company suing to get PickupPal.com shut down). One possible solution is to plan for such problems in advance, in order to be ready for them.

In real life, this strategy works surprisingly poorly. Anyone creating a new opportunity for social action has to understand the limits of planning: planning cannot completely substitute for experience. Because plans can go awry in so many ways and because users never behave the way you expect them to, the number of potential problems is almost unlimited. However, defending in advance against all imaginable problems will make things complicated for the users and hard to maintain; at the extreme case, preventing all possible misuse prevents all possible use as well. Even if someone did somehow defend in advance against all imaginable problems, they would still face the unimaginable problems.

As Brewster Kahle, a serial technology entrepreneur, once

said, "If you want to solve hard problems, have hard problems." Defending yourself in advance against the possible ramifications of success has strong diminishing returns. As a general rule, it is more important to try something new, and work on the problems as they arise, than to figure out a way to do something new without having any problems.

—*Clarity Is Violence*

We know quite a bit about taking advantage of new forms of participation, so why not just spell out for people what the service will do if they join and tell them the rules of the road? Instead of modeling positive deviance, which is a slow and painful way to change a culture, why didn't the Responsible Citizens just start with a contract? "We are going to clean up market streets in Lahore, and in doing so make Pakistan a better place. By signing here, you agree to show up in Anarkali and clean from 10 A.M. to 2 P.M. on Saturday."

Put that way, it's pretty easy to see why building something out of people's intrinsic motivations plus their free time is slow and uncertain work. Culture can't be created by fiat. (Little in the domain of cognitive surplus can be done by fiat.) But the task isn't just to get something done, it's to create an environment in which people want to do it. Working groups tend to accrue more governance as they grow, in part because the larger a group is, the more tension there can be between any two members of the group, and the greater the power imbalance between any member and the group as a whole. Even communities that end up with a lot of rules and requirements don't start out with them. Solving the problems as they arise means not putting a process in place until you need it.

David Weinberger, a fellow at Harvard's Berkman Center for Internet and Society, summed this point up nicely in a 2004 speech about groups and governance: *clarity is violence*. To use a historical analogy, the United States was founded in 1776, but the country that today's U.S. citizens actually live in was founded in 1787, the year the second (and current) constitution was written. The first constitution was written when the original thirteen olonies couldn't imagine giving up much of their sovereignty to participate in the larger federation of states, so the country in the 1770s was less a nation than a loose collection of competing entities.

By the late 1780s, the lack of mutual obligation was clearly keeping the union weak, so a new constitution was drawn up, obliging the states to contribute to national defense and forbidding them from erecting trade barriers, to name just two of the many new constraints. That constitution worked, and though it has been modified many times in the two centuries since it was ratified, the continuity between then and now is unbroken. For all the value of the 1787 constitution, though, it couldn't have been enacted in 1777, because the states wouldn't have been willing to yoke themselves to one another that tightly without an additional decade of experience. Groups tolerate governance, which is by definition a set of restrictions, only after enough value has accumulated to make the burden worthwhile. Since that value builds up only over time, the burden of the rules has to follow, not lead.

—Try Anything. Try Everything.

The Elements of Style (popularly known as Strunk and White) is a slim book that lays out rules for clear writing. But at the end of the book is this observation: the best writers sometimes disregard the rules. Abraham Lincoln's address commemorating the Battle of Gettysburg, a marvel of clarity and brevity, famously begins "Four score and seven years ago." This construction was archaic even then, but it was clearly in keeping with Lincoln's design for the speech.

The same tension between rules and design exists with social media; the rules are really guardrails for services, meant to illustrate the characteristic constraints and opportunities, but working social services also have an internal logic that matters more than any given prescription. Because the social tools we now have can shape public speech and civic action, people who design and use them have joined the experimental wing of political philosophy. The range of opportunities we can create for one another is so large, and so different from what life, until recently, was like, that no one person or group and no one set of rules or guides can describe all the possible cases. The single greatest predictor of how much value we get out of our cognitive surplus is how much we allow and encourage one another to experiment, because the only group that can try everything is everybody.

THREE WAYS TO MANAGE A REVOLUTION

When a new technology arrives, it has to get integrated into society somehow. It might be something minor (cheap long-distance calls, faster fax machines) or something major (the printing press, telephones). Major new possibilities always create some restructuring of society, because both the arrival of a new way to communicate and the ending of old constraints alter our connective tissue. The greater the difference between old and new possibilities, the less likely it is that the older behaviors will persist unaltered. Organizations that relied on uncontested access to public speech or coordinated action won't vanish, but competition from amateur and unmanaged groups will alter their relative importance. The open question for society becomes how to manage the social changes, and even upheavals, that come with such new possibilities.

When the telephone arrived, some feared it would lead to less formality between the sexes because women could converse with men to whom they had not been properly introduced. This fear of social chaos was correct. The phone did indeed help usher in significantly less formal dealings between men and women (a change later amplified by the arrival of that rolling boudoir, the automobile). It's hard to imagine what the average Victorian would make of the nature of relations between the sexes today.

Seeing what happened to society during earlier communications revolutions—the printing press, the telegraph, the mobile phone—we can ask: What *should* happen? What is the ideal way for a new technology to be integrated into society?

Let's divide this problem into a few different scenarios. One would be "As Much Chaos as We Can Stand": we let any would-be revolutionary try anything they like with the new technology, without regard for existing cultural or social norms or potential damage to current social institutions. Another scenario would be "Traditionalist Approval": the fate of any new technology would be put in the hands of the people responsible for the current way of doing things. It would be like leaving it up to the monks to decide how to use the printing press or to the post office to decide what to do with e-mail. A third scenario—call it "Negotiated Transition"—assumes a balanced conversation between radicals and traditionalists: radicals can propose uses of the new technology, then negotiate with traditionalists about how to take advantage of the new while preserving the best of the old.

Laid out this starkly, the third choice seems optimal. I now want to try to convince you that the right answer is actually the first one, "As Much Chaos as We Can Stand."

The middle scenario, Traditionalist Approval, would plainly be a disaster for society; if the beneficiaries of the old technology have veto power, they would effectively kill innovation, although not always out of self-interest. Someone who works at the post office may genuinely believe that the ways handwritten letters are superior to e-mail matter more than the ways e-mail is superior to letters. Such a person would, out of a deep and real conviction, want to limit the use of e-mail in order to preserve the older form of value, just as, a hundred years ago, manufacturers of horse-drawn carriages objected to Henry Ford's horseless variety on the grounds (again correct) that automobiles are far more dangerous than horses.

Bias in favor of existing systems is good, at least in periods of technological stability. When someone runs a bookstore, or a newspaper, or a TV station, it's advantageous to have those people think of their work as being critical for society, even if it isn't. This sort of commitment is good for morale and leads people to defend useful and valuable institutions.

However, that intellectual asset turns into a liability in times of revolution precisely because those deeply committed to old solutions cannot see how society would benefit from an approach incompatible with older models. Paradoxically, as we have seen, people committed to solving a particular problem also commit themselves to maintaining that problem in order to keep their solution viable. We can't ask people running traditional systems to evaluate a new technology for its radical benefits; people committed to keeping the current system will tend, as a group, to have trouble seeing value in anything disruptive.

Meanwhile, even in the "As Much Chaos as We Can Stand" scenario, the radicals won't be able to create any more change than the members of society can imagine. We've had the internet for forty years now, but Twitter and YouTube are less than five years old, not because the technology wasn't in place earlier but because society wasn't yet ready to take advantage of those opportunities. The upper limit of "As Much Chaos as We Can Stand" is thus the time and energy required for social diffusion. New ideas tend to spread slowly along social pathways; social diffusion isn't just about elapsed time but about the ways culture affects the use of new ideas. The embrace of social tools always contains surprises; Ushahidi.com was invented to track violence but later used to monitor elections; Wikipedia was designed as an encyclopedia but has also become critical for breaking global news. Questions of culture and context apply to the diffusion of all technology to some degree but to communications technology especially, since connective tissue varies with the kind of society being connected, and the kind of society being connected varies with its connective tissue.

The radicals will be unable to correctly predict the eventual ramifications because they have an incentive to overstate the new system's imagined value and because they will lack the capacity to imagine the other uses to which the tools will be put. That kills the "Negotiated Transition" scenario as well.

Proponents of the new and defenders of the old can't merely discuss the transition, because each group has systematic biases that make its overall vision untrustworthy; radicals and traditionalists start from different assumptions and usually end up talking past each other. The actual negotiated transition can happen only by letting the radicals try everything, because given their inability to predict what will happen, and given the natural braking functions of social diffusion, most of it will fail. The negotiation that matters isn't between radicals and traditionalists; instead it has to be with the citizens of the larger society, the only group who can legitimately decide how they want to live, given the new range of possibilities.

LOOKING FOR THE MOUSE

Our media environment (that is to say our connective tissue) has shifted. In a historical eyeblink, we have gone from a world with two different models of media—public broadcasts by professionals and private conversations between pairs of people—to a world where public and private media blend together, where professional and amateur production blur, and where voluntary public participation has moved from nonexistent to fundamental. This was a big deal even when digital networks were used only by an elite group of affluent citizens, but it's becoming a much bigger deal as the connected population has spread globally and crossed into the billions.

The world's people, and the connections among us, provide the raw material for cognitive surplus. The technology will continue to improve, and the population will continue to grow, but change in the direction of more participation has already happened. What matters most now is our imaginations. The opportunity before us, individually and collectively, is enormous; what we do with it will be determined largely by how well we are able to imagine and reward public creativity, participation, and sharing.

For those of us over forty, exercising this kind of imagination requires conscious effort, because it's so different from what we grew up with. At NYU, where I teach, I get to see the world through my students' eyes by listening to them talk, reading what they write, and watching what they do. This gives me some sense of the world as twenty-five-year-olds see it, and it looks very different from (and mostly better than) the world I grew up in. But the potential for really radical change may be even better illustrated through the eyes of children.

I was having dinner with a group of friends, talking about our kids, and one of them told a story about watching a DVD with his four-year-old daughter. In the middle of the movie, apropos of nothing, she jumped up off the couch and ran around behind the screen. My friend thought she wanted to see if the people in the movie were really back there. But that wasn't what she was up to. She started rooting around in the cables behind the screen. Her dad asked, "What you doing?" And she stuck her head out from behind the screen and said, "Looking for the mouse."

Here's something four-year-olds know: a screen without a mouse is missing something. Here's something else they know:

media that's targeted at you but doesn't include you may not be worth sitting still for. Those things make me believe that the kind of participation we're seeing today, in a relative handful of examples, is going to spread everywhere and to become the backbone of assumptions about how our culture should work. Four-year-olds, old enough to start absorbing the culture they live in but with little awareness of its antecedents, will not have to waste their time later trying to unlearn the lessons of a childhood spent watching *Gilligan's Island*. They will just assume that media includes the possibilities of consuming, producing, and sharing side by side, and that those possibilities are open to everyone. How else would you do it?

The girl's explanation has become my motto for what we might imagine from our newly connected world: we're looking for the mouse. We look everywhere a reader or a viewer or a patient or a citizen has been locked out of creating and sharing, or has been served up passive or canned experience, and we're asking. *If we carve out a little bit of the cognitive surplus and deploy it here, could we make a good thing happen?* I'm betting the answer is yes, or could be yes, if we give one another the opportunity to participate and reward one another for trying.

Acknowledgments

This book exists because Jennifer Pahlka wouldn't take No for an answer, insisting on my giving a talk about something new, just as I had finished the book tour for *Here Comes Everybody;* the framing of *Cognitive Surplus* was the result, so thank you, Jennifer.

The community at the Interactive Telecommunications Program at NYU has provided an incredible home, for me and for this work. Red Burns, the founder, to whom this book is dedicated; Dan O'Sullivan, the associate director; and my colleagues Tom Igoe, Nancy Hechinger, Nick Bilton, Kevin Slavin, and Kio Stark offered vital comments and support. I must also thank current and former students who have always asked sharp questions and pushed for clear answers, especially Cody Brown, Cheryl Furjanic, Jessica Hammer, John Geraci, Jorge Just, Liesje Hodgson, Steven Lehrburger, and Thomas Robertson. My research assistants at ITP, John Dimatos and Corey Menscher, have also been vital sources of observation about social media.

Chris Anderson, Lili Cheng, Tim O'Reilly, Andrew Stolli, and Kevin Werbach all provided their own observations, as well as offered public platforms for the development of this work. Long-running conversations with many colleagues have provided material and insights for this book, including Sunny Bates, Yochai

Benkler, danah boyd, Caterina Fake, Scott Heiferman, Tom Hennes, Liz Lawley, Beth Noveck, Danny O'Brien, Paul Resnick, Linda Stone, Martin Wattenberg, David Weinberger, and Ethan Zuckerman.

My agent, John Brockman, helped me clarify what I wanted to say, and Eamon Dolan and Helen Conford of Penguin Press helped me say it. Mel Blake, Ana Dane, Chris Meyer, and Vanessa Mobley all provided useful feedback on earlier versions, and Amy Lang was an invaluable research assistant.

Finally, of course, is Almaz, my endlessly patient wife, and Leo and Marina, my periodically patient children, who have been a source of inspiration and support throughout. Thank you all.

Notes

Chapter 1: Gin, Television and Cognitive Surplus

5 **TV quickly took up the largest chunk of our free time:** There are many sources tracking the number of hours of television use; although the hours vary somewhat by country, in the developed world the numbers range from the high teens to the high twenties. One interesting source for hourly figures, along with analysis, is "The Effects of Television Consumption on Social Perceptions: The Use of Priming Procedures to Investigate Psychological Processes," by L. J. Shrum, Robert S. Wyer, Jr., and Thomas C. O'Guinn, *The Journal of Consumer Research* 24.4 (1998): 447.

6 **Someone born in 1960 has watched something like fifty thousand hours of TV already:** This is simple extrapolation: at around twenty hours a week, someone who has grown up with television has seen about a thousand hours a year for every hour of their life. Another version of the same observation comes from Robert Kubey's *Television and the Quality of Life* ". . . a typical American would spend more than 7 full years watching television out of the approximately 47 *waking* years each of us lives by age 70."

6 **not only do unhappy people watch considerably more TV:** The full reference is Bruno Frey, Christine Benesch, and Alois Stutzer, "Does Watching TV Make Us Happy?" *Journal of Economic Psychology* 28.3 (2007): 283–313.

7 **Television viewing has come to displace:** Jib Fowles's book is *Why Viewers Watch: A Reappraisal of Television's Effects* (Newbury Park, CA: SAGE Publications, 1992), 37.

7 **people turn to favored programs when they are feeling lonely:** The full reference to Jaye Derrick and Shira Gabriel's study is "Social Surrogacy: How Favored Television Programs Provide the Experience of Belonging," *Journal of Experimental Social Psychology* 45.2 (2009): 352–62.

8 **television viewing plays a key role in crowding-out:** Luigino Bruni and Luca Stanca's 2008 paper is "Watching Alone: Relational Goods, Television and Happiness," *Journal of Economic Behavior & Organization* 65.3-4 (2008): 506–28.

8 **television can play a significant role in raising people's materialism:** In keeping with the expansion of economics to take on many other kinds of measurable

social issues, much of the most interesting work on television consumption is now being done by economists, including Marco Gui and Luca Stanca. Their working paper, referenced here, is "Television Viewing, Satisfaction and Happiness: Facts and Fictions," *University of Milan–Biocca, Department of Economics Working Paper Series,* 167 (2009), http://dipeco.economia.unimib.it/repec/pdf/mibwpaper167.pdf (accessed January 6, 2010).

9 **Back in 2006, Pluto was getting kicked out of the planet club:** Pluto has an orbit so uncharacteristic of the sun's other eight planets that after much discussion, the International Astronomical Union decided it was not to be labeled a planet anymore. There was much discussion of the decision, before and after; a good review of the decision itself is Mason Inman's "Pluto Not a Planet, Astronomers Rule," *National Geographic,* August 24, 2006, http://news.nationalgeographic.com/news/2006/08/060824-pluto-planet.html (accessed January 6, 2010).

10 **Martin Wattenberg, an IBM researcher:** Personal communication with author, April 2008.

11 **some cohorts of young people are watching TV less than their elders:** Paul Bond, "Study: Young People Watch Less TV," *Hollywood Reporter,* December 17, 2008, http://www.hollywoodreporter.com/hr/content_display/news/e3ic41d147829e712a6a6ecd990ea3a349c (accessed January 7, 2010).

11 **Young populations with access to fast, interactive media:** Marie-Louise Mares and Emory H. Woodard, "In Search of the Older Audience: Adult Age Differences in Television Viewing," *Journal of Broadcasting and Electronic Media* 50.4 (2006): 595–614.

11 **As Dan Hill noted:** Dan Hill's terrific essay, "Why *Lost* Is Genuinely New Media," was published on his blog, City of Sound, on March 27, 2006, http://www.cityofsound.com/blog/2006/03/why_lost_is_gen.html (accessed January 6, 2010).

12 **Charlie Leadbeater, the U.K. scholar of collaborative work:** Private communication with author, December 2009.

12 **But one of them, Gerald Berstell, chose to ignore the shakes:** Clayton M. Christensen, Scott D. Anthony, Gerald Berstell, and Denise Nitterhouse, "Finding the Right Job for Your Product," *MIT Sloan Management Review* 48.3 (2007): 38–47.

15 **In December 2007 a disputed election pitted supporters and opponents:** The results of the 2007 Kenyan election were widely discussed. A good contemporaneous description is Jeffrey Gettleman's "Disputed Vote Plunges Kenya into Bloodshed," *The New York Times,* December 31, 2007, http://www.nytimes.com/2007/12/31/world/africa/31kenya.html (accessed January 6, 2010).

15 **Ory Okolloh:** Okollah's role in founding Ushahidi is described by Dorcas Komo in "Kenyan Techie Honored for Role in Tracking Post-Election Violence," *Mshale: The African Community Newspaper,* July 3, 2008, http://mshale.com/article.cfm?articleID=18192 (accessed January 6, 2010).

16 **Ushahidi had been better at reporting acts of violence:** The Harvard study was written by Patrick Meier and Kate Brodock, "Crisis Mapping Kenya's Election Violence: Comparing Mainstream News, Citizen Journalism, and Ushahidi," Harvard Humanitarian Initiative, October 23, 2008, http://irevolution.word

press.com/2008/10/23/mapping-kenyas-election-violence (accessed January 6, 2010).

21 **an essay in 1997 called "Romancing the Looky-Loos":** Dave Hickey's marvelous collection of essays, including "Romancing the Looky-Loos," is *Air Guitar: Essays on Art and Democracy* (West Hollywood, CA: Foundation for Advanced Critical Studies, 1997): 146–54.

23 **In 2010 the global internet-connected population will cross two billion people:** There are many sources for predictions of growth of internet and mobile phone use. Two good ones are Dave Bailey's "Global Internet Population to Hit 2.2. Billion by 2013," *Computing,* July 21, 2009, http://www.computing.co.uk/computing/news/2246433/analyst-online-user-increase and Kirstin Ridley's "Global Mobile Phone Use to Pass 3 Billion," Reuters, June 27, 2007, http://uk.reuters.com/article/idUKL2712199720070627 (both accessed January 7, 2010).

24 **being part of a globally interconnected group:** World population estimate from Population Reference Bureau, *2009 World Population Data Sheet* (Washington, D.C.: PRB, 2009): 3, http://www.prb.org/pdf09/09wpds_eng.pdf, and age distribution estimate from the Central Intelligence Agency's "The World Factbook," https://www.cia.gov/library/publications/the-world-factbook/geos/xx.html#People (both accessed January 7, 2010).

25 **More is different:** Anderson's seminal essay, a touchstone for the ways in which aggregates of things exhibit new behaviors, is "More Is Different," from *Science* 177 (1972): 393–96.

25 **rose from a few million worldwide in 2000 to well over a billion today:** The mobile phone growth estimate is from the MIT Media Lab, in "Camera Culture," http://cameraculture.media.mit.edu (accessed January 7, 2010).

Chapter 2: Means

31 **Korean citizens staged public protests:** A good account of the Seoul protests as they were going on is Elise Yoon's "More Anti-Lee Myung-Bak Protests Continue," *The Seoul Times,* May 11, 2008, http://theseoultimes.com/ST/?url=/ST/db/read.php?idx=6585 (accessed January 7, 2010).

33 **I'm here because of Dong Bang Shin Ki:** Mizuki (Mimi) Ito, "Media Literacy and Social Action in a Post-Pokemon World" (paper presented as keynote address for the fifty-first NFAIS (National Federation of Advanced Information Services) annual conference, Philadelphia, PA, February 22–24, 2009), http://www.itofisher.com/mito/publications/media_literacy.html (accessed January 7, 2010)

34 **access to better, faster, and more widely available communications networks:** For a review of the various capabilities offered to citizens in high-tech cities, see "Tech Capitals of the World," *The Age,* June 18, 2007, http://www.theage.com.au/news/technology/tech-capitals-of-the world/2007/06/16/1181414598292.html (accessed January 7, 2010).

35 **during the month of May, that figure plummeted to less than 20 percent:** "No Bottom to Lee Myung-bak's Approval Ratings," *Anti2mb,* June 3, 2008, http://anti2mb.wordpress.com/2008/06/03/no-bottom-to-lee-myung-baks-approval-ratings (accessed January 7, 2010).

35 **websites were filled with images of policemen with water cannons:** You can find many of these videos on YouTube, such as "Seoul Protest Against Mad-Cow Beef," uploaded by a user going by dawitjaidii, at http://www.youtube.com/watch?v=mf-nutNE_iQ# (accessed January 7, 2010), or a trio of videos on the situation uploaded by a user going by digitallatlive at http://www.youtube.com/user/digitallatlive (accessed January 7, 2010). Interestingly, many of the videos are from users who created their YouTube accounts in early June 2008 and uploaded only one or a few protest videos, suggesting that the protests didn't just rely on social media, but further drove its use.

36 **The People Formerly Known as the Audience:** Jay Rosen has been using that phrase for much of this decade, but the most coherent statement of purpose is his blog post of that title, at http://journalism.nyu.edu/pubzone/weblogs/pressthink/2006/06/27/ppl_frmr.html (accessed January 8, 2010).

37 **trying to require citizens to use their real names online:** Michael Fitzpatrick, "South Korea Wants to Gag the Noisy Internet Rabble," *The Guardian*, October 8, 2008, http://www.guardian.co.uk/technology/2008/oct/09/news.internet (accessed January 8, 2010).

38 **As Ito describes the protesters:** Ito made these observations in a keynote speech, "Media Literacy and Social Action in a Post-Pokemon World," delivered to the fifty-first NFAIS annual conference. A rough transcript of the address is at http://www.itofisher.com/mito/publications/media_literacy.html (accessed January 8, 2010).

40 **hired a private detective to use PickupPal:** The detective's affidavit is at http://www.pickuppal.com/save/blog/res/PrivateInvestigationAffidavit.pdf.

40 **Trentway-Wagar invoked Section 11 of the Ontario Public Vehicles Act:** Daniel Goldbloom has a nice discussion of the legal situation of PickupPal in Ontario in "National Post Editorial Board on PickupPal: Carpooling Is Green and Cheap. So Why Is It Illegal in Ontario?" *National Post*, August 21, 2008, http://network.nationalpost.com/np/blogs/fullcomment/archive/2008/08/21/national-post-editorial-board-on-pickuppal-carpooling-is-green-and-cheap-so-why-is-it-illegal-in-ontario.aspx (accessed January 8, 2010).

41 **the Ontario legislature amended the Public Vehicles Act:** The website for the Save PickupPal movement put up a post after the legislative change: "Bill 118 Receives Royal Assent (We Won!)," Save PickupPal in Ontario, April 24, 2008, http://save.pickuppal.com/?p=16 (accessed January 8, 2010).

43 **a scribe could produce a single copy of a five-hundred-page book:** Paul Oskar Kristeller, *Studies in Renaissance Thought and Letters* (Rome, Italy: Ed. di Storia e Letteratura, 1993): 141.

43 **four ways a person could make books:** David Finkelstein and Alistair McCleery, *An Introduction to Book History* (London: Routledge, 2005): 68.

46 **All I had to do was type, then click a button marked "Publish":** Motoko Rich, a book reviewer for *The New York Times*, discusses the National Book Awards and Kingston's remarks in her blog post at the *Times:* "National Book Awards: Maxine Hong Kingston 2.0," *The New York Times*, November 20, 2008, http://papercuts.blogs.nytimes.com/2008/11/20/national-book-awards-maxine-hong-kingston-20 (accessed January 8, 2010).

47 **The multitude of books is a great evil:** William Hazlitt, ed. and trans., *The Table Talk of Martin Luther* (London: George Bell and Sons, 1902): 369.

47 **multiplication of books in every branch of knowledge is one of the greatest evils:** Chester Noyes Greenough, *The Works of Edgar Allan Poe: Volumes VII and VIII* (New York: Hearst's International Library Co., 1914): 164.

48 **The essay, titled "The Terrible Bargain We Have Regretfully Struck":** Melissa McEwan's, "The Terrible Bargain We Have Regretfully Struck," Shakesville from August 14, 2009 is at http://shakespearessister.blogspot.com/2009/08/terrible-bargain-we-have-regretfully.html. The comment thread is also extraordinary (accessed January 8, 2010).

50 **Whether this revolution in the reading habits of the American public:** Quoted in Kenneth Davis and Joann Giusto-Davis, *Two-Bit Culture: The Paperbacking of America* (New York and Boston: Houghton Mifflin, 1984): 68.

57 **The writer Nicholas Carr has dubbed this pattern digital sharecropping:** Nicholas Carr writes at his blog, Rough Type. "Sharecropping the Long Tail" is from December 19, 2006, http://www.roughtype.com/archives/2006/12/share cropping_t.php (accessed January 8, 2010).

59 **sued AOL on behalf of the ten thousand or so other volunteers:** Lisa Napoli covered the AOL lawsuit for *The New York Times:* "Former Volunteers Sue AOL, Seeking Back Pay for Work," *The New York Times,* March 26, 1999, http://www.nytimes.com/1999/05/26/nyregion/former-volunteers-sue-aol-seeking-back-pay-for-work.html? (accessed January 8, 2010).

59 **trying to make a dollar off the back of free slave labor:** Brian McWilliams, "AOL Volunteers Sue for Back Wages," *Internet News,* May 26, 1999, http://www.internetnews.com/xSP/article.php/8_127431. The site for the class action itself is at http://www.aolclassaction.com, and as of March 4, 2010, official notification of the class action had been mailed to all AOL community leaders.

61 **William Safire, an opinion columnist for *The New York Times*:** William Safire made these remarks in "What Else Are We Missing?" *The New York Times,* June 6, 2002, http://www.nytimes.com/2002/06/06/opinion/06SAFI.html? (accessed January 8, 2010).

62 ***"Skyful of Lies" and Black Swans*:** Nik Gowing, *"Skyful of Lies" and Black Swans: The New Tyranny of Shifting Information Power in Crisis* (Oxford: Reuters Institute for the Study of Journalism, 2009): 45–46, available as a PDF via http://reutersinstitute.politics.ox.ac.uk/publications/skyful-of-lies-black-swans.html (accessed January 15, 2010).

Chapter 3: Motive

66 **worked to make charitable giving part of life in Grobania:** "Grobanites for Charity: About Us," http://www.grobanitesforcharity.org/about (accessed January 8, 2010).

70 **a remarkably simple experiment that ignited a controversy:** Edward L. Deci, "Intrinsic Motivation, Extrinsic Reinforcement, and Inequity," *Journal of Personality and Social Psychology* 22.1 (1972): 113–20.

73 **asked whether they would approve a hypothetical government proposal:**

Bruno S. Frey, *Inspiring Economics: Human Motivations in Political Economy* (Cheltenham, England: Edward Elgar Publishing Limited, 2001): 77–81.

73 **where money was offered as a reward for volunteering:** Bruno S. Frey and Lorenz Goette, "Does Pay Motivate Volunteers?" (Zuerichbergstrasse, Zurich: Institute for Empirical Research in Economics, 1999), http://ideas.repec.org/s/zur/iewwpx.html.

73 **this sort of crowding out can appear in children as young as fourteen months:** Tomasello's research on children and their view of how things should be, by some ethical compass (a trait called "normativity," or the understanding and abiding by norms), was published as "The Sources of Normativity: Young Children's Awareness of the Normative Structure of Games," with his coauthors, H. Rakoczy and F. Wameken, in *Developmental Psychology* 44.3 (2008): 875–81.

74 **dozens of studies that had paid experimental subjects:** Judy Cameron and David Pierce, "Reinforcement, Reward, and Intrinsic Motivation: A Meta-Analysis," *Review of Educational Research* 64.3 (1994): 363–423.

74 **people were more motivated to do uninteresting tasks if you paid them:** Edward L., Deci, Richard Koestner, and Richard Ryan, "A Meta-Analytic Review of Experiments Examining the Effects of Extrinsic Rewards on Intrinsic Motivation," *Psychological Bulletin* 125.6 (1999): 627–68.

74 **crowding out of free choice can occur with the introduction of extrinsic motivations:** J. Cameron, K. M. Banko, and W. D. Pierce, "Pervasive Negative Effects of Rewards on Intrinsic Motivation: The Myth Continues," *Behavior Analyst* 24 (2001): 1–44.

75 **philanthropies that use 40 percent of donated money for expenses:** The American Institute of Philanthropy, "How American Institute of Philanthropy Rates Charities," http://www.charitywatch.org/criteria.html (accessed January 9, 2010).

76 **not graphics and gore but the feelings of control and competence:** Laura Sanders, "Gamers Crave Control and Competence, Not Carnage," *Science News* 175.4 (2009): 14.

78 **Commons-Based Peer Production and Virtue:** Benklar and Nissenbaum's paper, "Commons-Based Peer Production and Virtue," appeared in *The Journal of Political Philosophy* 14.4 (2006): 394–419.

79 **growth in postpartum support groups organizing via the internet:** Katherine Stone noted the prevalence of postpartum groups in "Postpartum Among Top 10 Fastest Growing Topics at Meetup.com," *Postpartum Progress,* October 8, 2009, http://postpartumprogress.typepad.com/weblog/2009/10/postpartum-among-top-10-fastest-growing-topics-at-meetupcom.html (accessed January 9, 2010).

80 **And then there's the section called Thank You:** The full name of the Thank You page is "Grobanites for Charity—A Special Thank You!" is at http://www.grobanitesforcharity.org/ty (accessed January 9, 2010).

84 **When Linus Torvalds first asked for help:** Torvald's original public announcement of what became Linux appeared as a question about a related operating system, Minix, on August 26, 1991, on the global discussion board usenet under the heading "What Would You Like to See Most in Minix?" Six other usenet users replied over the following two days. (http://groups.google.com/group/comp.os.minix/msg/b813d52cbc5a044b)

85 **Japanese anime (animated cartoons) are often subtitled in English:** Sean Leonard, "Celebrating Two Decades of Unlawful Progress: Fan Distribution, Proselytization Commons, and the Explosive Growth of Japanese Animation," *UCLA Entertainment Law Review* (Spring 2005): http://papers.ssrn.com/sol3/papers.cfm?abstract_id=696402 (accessed January 9, 2010).

85 **Yahoo.com hosts a mailing list for sufferers from Crohn's disease:** Yahoo! Health Groups, "Crohns: Living with Crohn's Disease, Yahoo! Groups, http://health.groups.yahoo.com/group/Crohns (accessed January 9, 2010).

85 **"EVERYTHING you need to know about the CPSIA":** Steve Spangler, "CPSIA Could Wage Severe Effects on Consumers, Retailers and the Economy," Steve Spangler Blog, January 3, 2009, http://www.stevespangler.com/in-the-news/cpsia-could-wage-severe-effects-on-consumers-retailers-and-the-economy (accessed January 9, 2010).

87 **postwar America saw a general decline in social connections:** Robert D. Putnam, *Bowling Alone: The Collapse and Revival of American Community* (New York: Simon & Schuster, 2000).

88 **the full crazy range of what people are actually interested in:** Nicholas Mirzoeff, personal communication with author, March 12, 2009.

88 **opinion pieces by a nerdy know-it-all named Larry Groznic:** Larry Groznic's columns can be found at *The Onion*, http://www.theonion.com/content/columnists/view/groznic (accessed January 9, 2010).

90 **a fanfic author with the pen name of Cassandra Claire:** Robert Covile covered the Cassandra Claire story as it was unfolding in "Boldly Go Where No One Has Gone Before," *Telegraph*, January 27, 2007, http://www.telegraph.co.uk/technology/3350729/Boldly-go-where-no-one-has-gone-before.html (accessed January 9, 2010).

90 **Some fan fiction writers even use a "legal" disclaimer:** Disclaimers are discussed in Rebecca Tushnet's "Copyright Law, Fan Practices, and the Rights of the Author," *Fandom: Identities and Communities in a Mediated World* (New York: New York University Press, 2009): 66. A search for the word "Disclaimer" in http://www.fanfiction.net/book/Harry_Potter will bring up many examples of the form.

91 **the other charge leveled at Cassandra Claire:** The Fan History Wiki has a discussion of this issue called "Cassandra Claire: Profiteering" at http://www.fanhistory.com/wiki/Cassandra_Claire#Profiteering (accessed January 9, 2010).

93 **twenty-four hours of video were being uploaded every minute onto YouTube:** M. G. Siegler, "Every Minute, Just About a Day's Worth of Video Is Now Uploaded to YouTube," *Tech Crunch*, May 20, 2009, http://www.techcrunch.com/2009/05/20/every-minute-just-about-a-days-worth-of-video-is-uploaded-to-youtube (accessed January 9, 2010).

93 **Twitter receives close to three hundred million words a day:** "In-depth Study of Twitter: How Much We Tweet, and When," *Royal Pingdom*, November 13, 2009, Pingdom AB, http://royal.pingdom.com/2009/11/13/in-depth-study-of-twitter-how-much-we-tweet-and-when (accessed January 9, 2010).

93 **asked its readers to rank a list of the 50 Most Beautiful People:** "Can You Trust Web 2.0?" *.net magazine*, April 4, 2008, Future Publishing, http://www.netmag.co.uk/zine/discover-culture/can-you-trust-web-2-0 (accessed January 9, 2010).

94 **The write-in campaign for Hank was started by Kevin Renzulli:** Joab Jackson covered the KOAM-inspired write-in at "Hanky-Panky," *Baltimore City Paper*, May 6, 1998, http://www.citypaper.com/columns/story.asp?id=5594 (accessed January 9, 2010).

CHAPTER 4: **Opportunity**

98 **stories with titles like "Old People Like the Internet":** Andy McCue, "Old People Like the Internet," *Silicon*, November 14, 2003, CBS Interactive Limited, http://www.silicon.com/technology/networks/2003/11/14/old-people-like-the-internet-39116903 (accessed January 9, 2010). Michael Agger, "Geezers Need Excitement: What Happens When Old People Go Online," *Slate*, September 11, 2008, http://www.slate.com/id/2199920/ (accessed January 9, 2010). Anne D'Innocenzio, "More Older People Turning to the Internet to Find Love," *redOrbit*, September 29, 2004, Associated Press, http://master.redorbit.com/news/technology/89595/more_older_people_turning_to_the_internet_to_find_love/index.html (accessed January 9, 2010).

99 **theory-induced blindness:** Personal communication with author, May 2009.

100 **energetic techies built a phone booth:** Brad Templeton, "The Phone Number Is Dead," Brad Ideas, October 1, 2005, http://ideas.4brad.com/node/269 (accessed January 9, 2010).

102 **the self-styled Z-Boys, began riding on skateboards inside the empty pools:** Regina Hackett, "Seattle Artists Roll Out Dynamic Skateboard Art Celebrating the Legendary Z Boys," *Seattle Post-Intelligencer*, February 3, 2008, http://www.seattlepi.com/visualart/107733_skateboard08.shtml (accessed January 9, 2010).

103 **lead user innovation:** Eric von Hippel, "Lead Users: A Source of Novel Product Concepts," *Management Science* 32.7 (1986): 791–805.

103 **their first formal skateboarding competition:** "The Z-BOY Story," Z-Boy, http://z-boy.com (accessed January 9, 2010).

104 **improve the ideas of a group and spread them:** Michael Farrell, *Collaborative Circles: Friendship Dynamics and Creative Work* (New York: New York University Press, 2001).

105 **the Ultimatum Game:** Werner Güth, Rolf Schmittberger, and Bernd Schwarze, "An Experimental Analysis of Ultimatum Bargaining," *Journal of Economic Behaviour and Organization* 3.4 (1982): 367–88.

107 **Versions were run with hundreds of dollars at stake:** Joseph Henrich, Robert Boyd, Samuel Bowles, Colin Camerer, Ernst Fehr, and Herbert Gintis, *Foundations of Human Sociality: Economic Experiments and Ethnogra-*

phic Evidence from Fifteen Small-Scale Societies (Oxford: Oxford University Press, 2004).

111 **In his book *Public Associations in Civil Life*:** Alexis de Tocqueville, "Chapter XXVII: Public Associations," *Democracy in America, Vol. 2* (New York: George Adlard, 1838): 593–607.

112 **we design systems that reward selfish people:** Elinor Ostrom, *Governing the Commons: The Evolution of Institutions for Common Action* (Cambridge, U.K.: Cambridge University Press, 1990).

113 **office workers will take fewer if there are paper cutouts of eyes:** Lee Bowman, "Office Workers Add to Coffee Kitty if Watched," Scripps Howard News Service, June 28, 2006, http://www.abqtrib.com/news/2006/jun/28/office-workers-add-to-coffee-kitty-if-watched (accessed January 7, 2010)

114 **Conscience is the little voice that tells you someone might be looking:** H. L. Mencken, *A Mencken Chrestomathy: His Own Selection of His Choicest Writings* (New York: Vintage, 1982): 617.

114 **Behlendorf was the primary programmer for Apache:** "Brian Behlendorf, Founding Member of the Apache Software Foundation Speaks on How Open Source Developers Can Save the World," Bitsource, October 6, 2009, The Bitsource, http://www.thebitsource.com/2009/10/06/brian-behlendorf-apachecon-keynote (accessed January 9, 2010).

116 **People do get paid to work on it:** John Naughten, "The High Tech Gift Culture," *A Brief History of the Future: Origins of the Internet* (New York: The Overlook Press, 2000); also at http://www.briefhistory.com/pages/extract4.htm (accessed January 9, 2010).

119 **upended the idea that humans always determine value rationally:** Dan Ariely, *Predictably Irrational: The Hidden Forces That Shape Our Decisions* (New York: Harper, 2008).

119 **commons-based peer production:** Yochai Benkler, "Coase's Penguin, or, Linux and the Nature of the Firm," *Yale Law Journal* 112 (2002): 371–99.

120 **Napster users could share a list of the songs:** Spencer E. Ante, "Napster's Shawn Fanning: The Teen Who Woke Up Web Music," *BusinessWeek*, April 12, 2000, Bloomberg, http://www.businessweek.com/ebiz/0004/em0412.htm (accessed January 9, 2010).

120 **Napster acquired tens of millions of users in less than two years :** Benny Evangelista, "News Analysis: Internet Music Will Still Play on Despite Napster's Uncertain Future," *San Francisco Chronicle*, February 18, 2001, Hearst Communications, http://www.sfgate.com/c/a/2001/02/18/BU39387.DTL (accessed January 9, 2010).

124 **"He who receives ideas from me":** Quoted in John Pitman, "Open Access to Professional Information," *IMS Bulletin* 36.8 (2007): 13.

125 **"Triumph of the Default":** Kevin Kelly, "Triumph of the Default," *The Technium*, June 22, 2009, Creative Commons, http://www.kk.org/thetechnium/archives/2009/06/triumph_of_the.php (accessed January 9, 2010).

126 **tired of their country's divisive politics:** Sabrina Tavernise, "Young Pakistanis Take One Problem into Their Own Hands," *The New York Times*, May 18, 2009, http://www.nytimes.com/2009/05/19/world/asia/19trash.html (accessed January 9, 2010). The Responsible Citizens site is at http://www.zimmedarshehri.com (accessed January 7, 2010).

128 **The Strength of Weak Ties:** Mark S. Granovetter, "The Strength of Weak Ties," *American Journal of Sociology* 78.6 (1973): 1360.

128 **social networks spread all kinds of behaviors:** Nicholas Christakis and James Fowler discuss "Social Networks and Happiness" in *Edge* (2008), http://www.edge.org/3rd_culture/christakis_fowler08/christakis_fowler08_index.html (accessed January 9, 2010). Their book is *Connected: The Surprising Power of Our Social Networks and How They Shape Our Lives* (New York: Little, Brown, 2009).

Chapter 5: Culture

131 **"A Fine Is a Price":** Uri Gneezy and Aldo Rustichini, "A Fine Is a Price," *Journal of Legal Studies* 29.1 (2000): 1–17.

137 **our Invisible College:** Richard Weld, *A History of the Royal Society* (London: John W. Parker, West Strand, 1848): 39.

138 **Hermetic Books have such involved Obscuritys:** Quoted by Lawrence Principe in "Boyle's Alchemical Pursuits," *Robert Boyle Reconsidered*, M. Hunter (ed.) (Cambridge, UK: Cambridge University Press,1994), 9.

140 **In his book *The Economics of Knowledge*:** Dominique Foray's book, *The Economics of Knowledge* (Cambridge, MA: MIT Press, 2004).

142 **a kite-sailing community called Zero Prestige:** Eric von Hippel, *Democratizing Innovation* (Cambridge, MA: MIT Press, 2005): 103–25.

143 **coined the term "communities of practice:** Etienne Wenger, *Communities of Practice: Learning, Meaning, and Identity* (Cambridge, U.K.: Cambridge University Press, 1998).

145 **Andrew McWilliams, his professor:** Andrew McWilliams's actions are reported in "Student Faces Facebook Consequences," *Toronto Star*, March 6, 2008, http://www.thestar.com/News/GTA/article/309855 (accessed January 9, 2010).

146 **if work is to be done individually and students collaborate, that's cheating:** James Norrie is quoted in "Facebook User Can Stay at Ryerson," *Toronto Star*, March 19, 2008, http://www.thestar.com/article/347688 (accessed January 9, 2010).

146 **If this is cheating, then so is tutoring:** Avenir is quoted in "Student 'Plagiarised' Via Facebook," *Times Higher Education*, March 20, 2008, TSL Education LTD., http://www.timeshighereducation.co.uk/story.asp?storyCode=401139§ioncode=26 (accessed January 9, 2010).

151 **an essay called "The Zagat Effect":** Steven Shaw's "The Zagat Effect" appeared in *Commentary Magazine* (November 2000): 47–50.

154 **Chris Anderson, author of *Free*:** Chris Anderson, *Free: The Future of a Radical Price* (New York: Hyperion, 2009): 194–95.

156 **The largest studies on PLS or PMA:** "Charting the Course of PLS and PMA," The PatientsLikeMe Blog, August 11, 2009, http://blog.patientslikeme.com/2009/08/11/charting-the-course-of-pls-and-pma (accessed January 9, 2010).

157 **he got his neurologist to alter his 10mg dose of baclofen:** Thomas Goetz tells the baclofen story in "Practicing Patients," about the rising involvement of patients in all aspects of their diagnosis and care. *The New York Times Magazine,* March 23, 2008, http://www.nytimes.com/2008/03/23/magazine/23patients-t.html (accessed January 9, 2010).

158 **it also has an "openness philosophy":** "The Value of Openness," The PatientsLikeMe Blog, December 13, 2007, http://blog.patientslikeme.com/2007/12/13/the-value-of-openness (accessed January 9, 2010).

CHAPTER 6: **Personal, Communal, Public, Civic**

161 **Steve Ballmer of Microsoft denounced the shared production of software:** Lea Graham, "MS Ballmer: Linux Is Communism," *The Register,* July 31, 2000, http://www.theregister.co.uk/2000/07/31/ms_ballmer_linux_is_communism/ (accessed January 10, 2010).

162 Robert McHenry, "The Faith-Based Encyclopedia," *Technology Commerce Society Daily,* November 15, 2004, http://www.tcsdaily.com/article.aspx?id=111504A (accessed January 10, 2010).

162 **compared bloggors to monkeys:** Andrew Keen, *The Cult of the Amatuer: How Blogs, MySpace, YouTube, and the Rest of Today's User-Generated Media Are Destroying Our Economy, Our Culture, and Our Values* (New York: Broadway Business, 2007): 2.

163 **a slim volume called *Experiences in Groups*:** W. R. Bion, *Experience in Groups and Other Papers* (New York: Routlege, 1991).

165 **The video starts simply enough:** "Couch Surfing," *Current TV,* July 21, 2007, http://current.com/items/76406002_couch-surfing.htm (accessed January 10, 2010).

167 **Hitchhiking is choosing to have faith in other human beings:** Pippa Bacca and Siliva Moro, "Progretto," *Brides on Tour,* http://bridesontour.fotoup.net/progetto.html (accessed January 10, 2010).

167 **Shortly after leaving Istanbul, Pippa Bacca was abducted:** Laura Kind, "A Plea for Peace in White Goes Dark," *Los Angeles Times,* May 31, 2008, http://articles.latimes.com/2008/may/31/world/fg-pippa31?pg=5 (accessed January 10, 2010).

169 **religious fundamentalists named Sri Ram Sene attacked women drinking:** "Young India Vents Anger Over Mangalore Incident on Internet," *Thaindian News,* January 27, 2009, http://www.thaindian.com/newsportal/uncat egorized/young-india-vens-anger-over-mangalore-incident-on-internet_100147756.html (accessed January 10, 2010).

169 **The founder of Sri Ram Sene, Pramod Muthali:** "Girls Assaulted at Mangalore Pub," *The Times of India,* January 26, 2009, http://timesofindia.indiatimes.com/Cities/Girls_assaulted_at_Mangalore_pub/articleshow/4029791.cms (accessed January 10, 2010).

169 **the Association of Pub-going, Loose and Forward Women:** Philip Reeves, "'Moral Police' in India to Get Valentine's Underwear," National Public Radio, February 13, 2009, http://www.npr.org/templates/story/story.php?storyId=100624625 (accessed January 10, 2010).

170 **Susan's campaign flooded Muthali's office with *chaddis*:** Robert Mackey, "Indian

Women Fight Violence with Facebook and Underwear," New York Times Lede Blog, February 13, 2009, http://thelede.blogs.nytimes.com/2009/02/13/indian-women-use-facebook-for-valentines-protest (accessed January 10, 2010).

171 **the state of Mangalore arrested Muthali:** "Muthali Arrested to Save V-Day in Karnataka," *Indian Express*, February 13, 2009, http://www.indianexpress.com/news/muthalik-arrested-to-save-vday-in-karnataka/423184 (accessed January 9, 2010).

175 **You can always get what you want . . .:** From Gary Kamiya's "The Death of the News," Salon, February 17, 2009, http://www.salon.com/opinion/kamiya/2009/02/17/newspapers/index.html (accessed January 10, 2010).

176 **In a free culture, you get what you celebrate:** Dean Kamen describes this idea in "You Get What You Celebrate," Xconomy Boston, January 2, 2008, http://www.xconomy.com/boston/2008/01/02/you-get-what-you-celebrate (accessed January 10, 2010).

177 **the assumption that "people are basically good":** "Pierre Omidyar on 'Connecting People,'" *BusinessWeek*, June 20, 2005, http://www.businessweek.com/magazine/content/05_25/b3938900.htm (accessed January 10, 2010).

177 **This idea, as noble as it is, didn't quite pan out for eBay:** Tobias J. Klein, Christian Lambertz, Giancarlo Spagnolo, and Konrad O. Stahl, "The Actual Structure of eBay's Feedback Mechanism and Early Evidence on the Effects of Recent Changes," *International Journal of Electronic Business* 7.3 (2009): 301–20.

177 **an 8 percent premium on price:** Paul Resnick published these findings with his coauthors Richard Zeckhauser, John Swanson, and Kate Lockwood, in "The Value of Reputation on eBay: A Controlled Experiment," *Experimental Economics* 9.2 (2006): 79–101.

179 **added a fake quote to composer Maurice Jarre's Wikipedia page:** Shawn Pogatchnik discussed Fitzgerald's actions in "Student Hoaxes World's Media on Wikipedia," MSNBC, May 12, 2009, http://www.msnbc.msn.com/id/30699302 (accessed January 10, 2010).

Chapter 7: Looking for the Mouse

185 **notes in his book *The Success of Open Source*:** Steven Weber, *The Success of Open Source* (Cambridge, MA: Harvard University Press, 2005): 272.

188 **He got a loan to enter the indulgence-printing business:** The British Library discusses Gutenberg's printing of indulgences in its documentation of Gutenberg's Bible: http://www.bl.uk/treasures/gutenberg/indulgences.html (accessed January 9, 2010).

188 **John Tetzel, the head pardoner for German territories:** Tetzel's place in history was largely secured by Martin Luther's objections to indulgences in 1517, but his name recently reappeared when the Catholic Church brought back indulgences in 2008; in discussing this change, John Allen references Tetzel's phrase in the Room for Debate blog, http://roomfordebate.blogs.nytimes.com/2009/02/13/sin-and-its-indulgences (accessed January 7, 2010).

190 **As Elizabeth Eisenstein notes in *The Printing Press as an Agent of Change*:** Elizabeth Eisenstein, *The Printing Press as an Agent of Change: Communications* *and Cultural Transformations in Early-Modern Europe* (Cambridge, U.K.: Cambridge University Press, 1980).

192 **a computer system called PLATO:** Elisabeth Van Meer discusses this history in "PLATO: From Computer-Based Education to Corporate Social Responsibility," *Iterations: An Interdisciplinary Journal of Software History* (2003): 6–22.

196 **"The behavior you're seeing is the behavior you've designed for":** Joshua Porter, "The Behavior You're Seeing Is the Behavior You've Designed For," Bokardo, July 28, 2009, http://bokardo.com/archives/the-behavior-youve-designed-for (accessed January 10, 2010).

203 **One of the most parsimonious examples of this pattern on the web is from JavaRanch:** "Be Nice," JavaRanch, http://faq.javaranch.com/java/BeNice (accessed January 10, 2010).

203 **it sometimes upgraded its software every half hour:** Nisan Gabbay, "Flickr Case Study: Still About Tech for Exit?" *Startup Review,* August 27, 2006, http://www.startup-review.com/blog/flickr-case-study-still-about-tech-for-exit.php (accessed January 10, 2010).

203 **has its users watch people trying to use their service every day:** Meetup's user-testing setup observed by the author and discussed at "Meetup's Dead Simple User Testing," http://www.boingboing.net/2008/12/13/meetups-dead-simple.html (accessed January 9, 2010).

205 **"If you want to solve hard problems, have hard problems":** Brewster Kahle advised on the Libarary of Congress's digital preservation efforts starting in 2003 (a project I also worked on); he made this remark at a meeting in Berkeley, California, in April 2003.

206 **clarity is violence:** David Weinberger made this observation in a talk called "What Groups Will Be," presented at the O'Reilly Emerging Technology Conference, Santa Clara, CA, April 26, 2003.

207 ***The Elements of Style*** **(popularly known as Strunk and White):** William Strunk's book *The Elements of Style* (Geneva, NY: Press of W. P. Humphrey, 1918) was later updated and expanded by E. B. White, hence the popular name.

208 **some feared it would lead to less formality between the sexes:** Claude S. Fisher, *America Calling: A Social History of the Telephone to 1940* (Berkeley, CA: University of California Press, 1994): 356.

Index

About the Author

Clay Shirky teaches at the Interactive Telecommunications Program at New York University, where he researches the interrelated effects of our social and technological networks. He has consulted with a variety of groups working on network design, including Nokia, the BBC, NewsCorp, Microsoft, BP, Global Business Network, the Library of Congress, the U.S. Navy, and Lego. His writings have appeared in *The New York Times, The Wall Street Journal, The Times* (London), *Harvard Business Review, Business 2.0,* and *Wired.*